PROPAGATION

PROPAGATION

John I. Wright

BLANDFORD PRESS
LONDON

Blandford Press

an imprint of
Cassell plc
Artillery House, Artillery Row
London SW1P 1RT

Copyright © John I. Wright 1985

Adapted from material first published in
Plant Propagation for the Amateur Gardener

First published 1985
Reprinted 1988

All rights reserved. No part of this book may
be reproduced or transmitted in any form or by
any means, electronic or mechanical, including
photocopying, recording or any information storage
and retrieval system, without permission in
writing from the Publisher.

British Library Cataloguing in Publication Data

Wright, John I.
  Propagation: how to grow plants from seeds,
  cuttings and other methods.
  1. Plant propagation
  I. Title
  635'.043   SB119

ISBN 0 7137 1611 8

Distributed in the United States by
Sterling Publishing Co., Inc.,
2 Park Avenue, New York, N.Y. 10016

Distributed in Australia by
Capricorn Link (Australia) Pty Ltd.,
PO Box 665, Lane Cove, N.S.W. 2066

Typeset by Megaron Typesetting, Bournemouth

Printed in Great Britain by Richard Clay Ltd., Chichester

CONTENTS

Introduction 6

1 Cuttings 7
Basal cuttings; Heel cuttings; Bud cuttings; Eye cuttings; Stem cuttings; Tip cuttings; Internodal and nodal cuttings; Irishman's cuttings; Leaf cuttings; Piping cuttings; Root cuttings; Suitable plants for this method

2 Seeds 24
Types of seeds and methods of encouraging germination; Collecting and preparing seed; F1 hybrids and hybridisation; Sowing seed; Sowing outdoors; Spores; Spawn; Suitable plants for this method

3 Division 46
Fibrous-rooted plants; Woody-rooted plants; Offsets; Rhizomes; Suckers; Tubers; Suitable plants for this method

4 Layering 56
Air layering; Serpentine layering; Tip layering; Suitable plants for this method

5 Grafting 62
Splice graft; Whip and tongue graft; Wedge or cleft graft; Saddle graft; Approach grafting; Framework grafting; Crown or rind grafting; Double-working; Flat grafting; Veneer grafting; Rootstocks; Suitable plants for this method

6 Budding 78
Chip budding; Rootstocks; Suitable plants for this method

7 Greenhouses, Frames and Propagators 88

Index 94

INTRODUCTION

This book sets out to give the amateur gardener, experienced or otherwise, a comprehensive account of the different methods of propagating his stock of plants, from the simplest form of division to the advanced techniques of grafting. The basic principles of each method are described in the main part of the chapters, and at the end of each chapter there is a table showing some of the most common plants to be propagated by that particular method. These tables are by no means exhaustive, especially in the cases of cuttings, seeds and division, but give a good indication of the range of plants involved.

By employing the techniques described in the following pages, the reader should be able to reduce his expenditure on new plants considerably, and also take great pride in the development of his or her garden, greenhouse or houseplant display.

1 CUTTINGS

Raising plants from cuttings is a quick, easy and cheap method of increasing the plants in your garden or greenhouse. Many kinds can be propagated in this way, and cuttings are so easy to take that any visit to a friend's garden can yield a few new specimens with no detrimental effect on either the parent plants or your friendship.

A cutting is any part of a plant which, when taken from its parent and treated in the correct manner, becomes an individual plant in its own right. Most cuttings are stem cuttings, but in some cases plants can also be propagated from their leaves and roots.

As the rooted cutting will be identical to its parent, great care should be taken when selecting that parent. A poor-growing, spindly plant is unlikely to produce a prize specimen from a cutting, and neither is a plant riddled with disease. Plants selected for cutting propagation should therefore be the strongest-growing, most healthy plants available.

Cuttings of some type can be taken throughout the year, the owner of a heated greenhouse quite naturally having the greatest choice of subject and propagating season; yet even without such facilities there is rarely a time of the year when something cannot be prepared as a cutting.

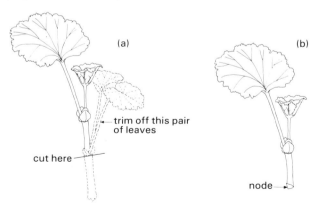

FIG. 1 (a) Softwood cutting from *Pelargonium* (geranium). (b) Cutting prepared for planting.

Almost all cuttings are prepared in the same way before planting. They should all be taken from the parent plant with a sharp knife or razorblade. Sharpness is essential as bruising caused by a blunt knife will possibly result in the cut end of the cutting rotting away and subsequently failing. For this same reason, cuttings should never be taken or prepared with the fingernails; although they may be broken off by bending in some cases, a knife only being used if any trimming of loose ends seems necessary.

Most stem cuttings should be taken immediately below a leaf-joint or node (*see* 'Internodal and Nodal Cuttings'), making the cut as straight as possible across the stem. Once removed from the parent plant any leaves, stalks or buds likely to be beneath the surface of the soil when the cutting is planted should be taken from as close to the stem as possible with the knife. When the cutting has been prepared in this manner it is ready for planting (Fig. 1).

To assist with root formation, the cutting is best dipped into a hormone rooting compound of a suitable type, first dipping it into a little clean water to assist the powder to stick to the stem. Although some people disagree, it has been my experience that saving the powders from one year to the next is false economy, as the effectiveness appears to wear off with exposure to the air. For my part I buy fresh stock in each season. Cuttings should be dipped into the powder to a depth in relation to their size; 6 mm or so for the smallest softwood to perhaps 4 cm for tall hardwood cuttings.

Leaf and root cuttings are prepared in a different manner, of course, and this is described later, but whichever method is used for taking or preparing a cutting a correctly formulated compost is of prime importance for the successful rooting of a large number of subjects.

There are a number of formulae for making up composts suitable for rooting cuttings, or 'striking' them, as the process is generally known. For most purposes the John Innes Seed Compost is ideal. This, along with proprietary seed composts, can be bought from garden centres etc, or it can be made up yourself from the separate ingredients.

The measurements of these various components is by volume, the basic unit being the bushel (36 dry litres or 8 dry gallons), or for larger quantities, the cubic metre (cubic yard). As most gardeners' requirements are likely to be in bushels rather than cubic metres it is convenient to have a container in which the former amount can be measured. Four 2-gallon buckets yield a bushel, but a rectangular wooden box specially made for the purpose is best. A box with an internal measurement of $56 \times 25 \times 25$ cm will hold 1 bushel of soil or compost. The following are some of the compost formulae suitable for cuttings.

John Innes Seed Compost

2 parts (by bulk) sterilised loam.

1 part moss peat.

1 part coarse sand or grit.

To each bushel is added 40 g 18% Superphosphate, 20 g chalk.

Soilless compost

You can buy peat-based proprietary composts, but if you use one of these choose a seed and cutting, or a 'universal' formula. It is easy to make your own soilless cutting compost, however, from silver sand (2 parts by bulk) and moss peat (1 part). As this mixture contains no nutrients, the rooted cuttings will have to be potted up without delay.

Various other materials can be used as the medium for raising cuttings, the most popular probably being the substance vermiculite. This is a type of mica found in the USA and South Africa which, when heated and expanded, becomes light and flaky and has a great capacity for retaining water without becoming waterlogged. As with other soilless composts it is essential to transplant or feed cuttings rooted in such material as soon as roots are formed. The vermiculite used for loft insulation may be too alkaline, so it is wise to choose 'horticultural vermiculite', which will have had the pH adjusted.

There is no special requirement as to the type of container suitable to hold the compost in which to root cuttings, the only real essential being that it has free drainage. Although wooden boxes, plastic trays and pots are all suitable it will be found that many cuttings root especially well if planted around the rim of a clay pot.

Watering cuttings can often prove a difficulty, as too much

water can cause the the cutting to rot below ground and too little will cause drying out. The best rule to follow is to start off with moist compost and give the cuttings a good watering with a fine rose immediately after planting, then leave them alone. If the weather is hot then spray them over lightly night and morning but do resist the temptation to overwater. Remember that many softwood cuttings will droop for the first few days after planting in any case, this being a natural reaction to being separated from the roots of the parent, and further watering would not have any effect on this.

Once cuttings are rooted they are transplanted, more often than not into individual 7.5 cm pots, or they may be planted out in boxes or in frames. Those transplanted into pots or boxes should be treated in a similar manner to seedlings, and details of

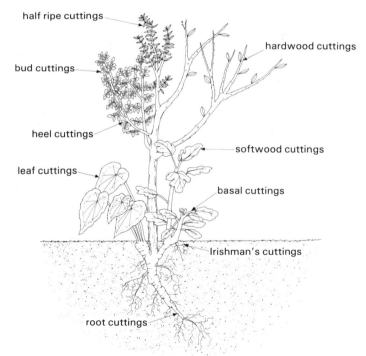

FIG. 2 Different types of cutting and their relative positions on the plant.

composts can be found in Chapter 2.

There are several different types of cuttings (Fig. 2), which for ease I will describe under their appropriate names.

BASAL CUTTINGS

A basal cutting is a young shoot cut or gently pulled from the base of a parent plant. Generally, such cuttings need little preparation other than the removal of a few lower leaves before planting. They should be taken when they are as young and soft as is convenient to handle. A cutting length 2.5-10 cm is most usual. They need some form of protection to avoid drying out while rooting, the heated propagating frame being best, especially early on in the season when most cuttings of this type are taken.

Many perennials and alpine plants can be propagated by means of basal cuttings.

HEEL CUTTINGS

A heel cutting is usually a sideshoot of hardwood or semi-hardwood pulled from the parent plant complete with a 'heel' of bark from the main stem (Fig. 3). Many shrubs such as *Jasminum nudiflorum* are found to root much better from such cuttings. They should be prepared and grown as ordinary hardwood and semi-hardwood subjects, but without doing anything to the heel at the base of the cutting other than trimming any loose ends with a sharp knife before planting.

Although perhaps more a basal than a heel cutting, it will be found that a dahlia cutting will root more readily if pulled or cut from the crown of the tuber with a little bark or rind attached. Certainly, the speed of rooting is much better in cuttings treated in this way.

BUD CUTTINGS

Taking a bud cutting (Fig. 3) is very similar to preparing a bud for budding, the main difference being that with the cutting the leaf is retained attached to the slice of stem.

A semi-hardwood shoot should be chosen to supply the bud, usually in the late summer from the same season's growth. A slice of stem around 2.5 cm in length is cut from the parent plant, each cutting containing a leaf and a leaf axil containing a bud. Plant the cutting so that the bud and leaf are just above the surface of

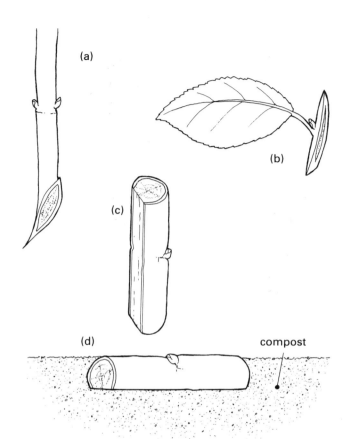

FIG. 3 (a) Heel cutting. (b) Bud cutting. (c) Eye cutting. (d) An eye cutting pressed into compost.

the compost and place in a frame or propagator with gentle bottom heat until well rooted.

Camellias and roses are among the plants which can be propagated by this method.

EYE CUTTINGS

This is a very successful method of propagation usually used for vines. An eye cutting is a hardwood stem cutting taken from the parent plant during the dormant periods of autumn and winter.

The best cutting is one about 4 cm long, containing one 'eye' or bud (Fig. 3). A length of vine stem can be cut into many sections each containing a bud. With the sections so cut, a strip of bark is removed from the opposite side of the stem to the bud, and the cutting pegged down on a loamy compost, J.I. Seed Compost being ideal.

With the cutting lightly pegged down, sprinkle silver sand liberally over the compost surface and press the cutting in so that only the bud protrudes above the sand. A bottom heat of 24°C is required for rooting, making a propagating frame of some kind essential.

With this temperature the compost is inclined to dry out rapidly, so it is a good plan to check periodically to make sure that it remains moist around the cutting.

STEM CUTTINGS

Most cuttings fall into this category. A stem cutting is any cutting taken from the main shoot of a plant or any sideshoots growing from the same plant or stem. The category is divided into three sections; softwood, semi-hardwood and hardwood cuttings. Particularly with shrubs and trees, cuttings from young plants are to be preferred to cuttings from very old ones as for some reason they seem to produce roots more easily.

Softwood cuttings

This is a group name given to any cutting taken from shoots that are still soft and have not become hard or 'woody'. Usually, the cutting is 4-7.5 cm long, and is prepared like other stem cuttings with the lower leaves removed and the stem cut immediately below a node with a razorblade or knife before planting.

As softwood cutting are by nature soft and fleshy, they are easily dried out and killed, so they have to be rooted in a close damp atmosphere, and need at least a little bottom heat to encourage root formation. A propagating frame with a temperature controlled at 16-18°C is ideal.

A sandy type of compost suits them best, and when planting it is a good rule not to plant too deeply; just enough to hold the new plant upright is sufficient (Fig. 4).

Many herbaceous plants as well as chrysanthemums, dahlias and geraniums are propagated from softwood cuttings.

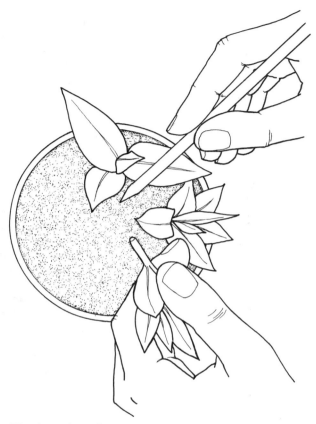

FIG. 4 Planting softwood cuttings of *Tradescantia* in a small pot.

Half-ripe or semi-hardwood cuttings

The term 'semi-hardwood' covers a number of types of cutting, but basically it means that the cutting is prepared from wood of the current season's growth. This is usually found on the sideshoots or on the top of the main stem of the parent plant in late summer.

Cuttings are usually 5 to 10 cm long, and are prepared by trimming the cutting with a straight cut below a node and removing a few lower leaves. Planting is best done in J.I. Seed Compost or a peat/sand mixture, the cuttings being inserted to a quarter of their length. A shaded cold frame is the best situation for rooting semi-hardwoods. Very many species of shrub and shrubby plants can be propagated from cuttings of this type (Fig. 5).

Hardwood cuttings

This type of cutting is ideal for many hardy shrubs, trees and soft fruits, and is one of the easiest to handle. The hardwood cutting is a stem cutting taken from the parent plant at the end of the growing season when the wood of the current season's growth has matured ready for the winter. October and November are the months when most of such cuttings are taken.

Cuttings of 25-30 cm should be cut from the parent plant just below a node, and any lower leaves still attached should be removed. Any well-drained and slightly shaded corner of the open garden will be suitable to root these cuttings. Planted out, they will remain dormant throughout the winter, breaking into growth the following spring. In colder districts it is better to plant them in the soil of a cold frame to give them extra protection, and this is what I would recommend.

Planting any quantity is easily done by making a V-shaped trench 10 cm or so deep with a spade. This is done by simply pushing the spade into the soil and waggling it back and forth a

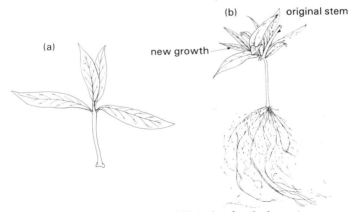

FIG. 5 (a) A half-ripe nodal cutting of *Weigela* taken in August.
(b) *Weigela* cutting rooted and ready for potting on in March after overwintering in a cold frame.

couple of times. In anything other than very light soils a small amount of silver sand is best spread along the bottom of the V to assist with drainage and encourage rooting. Place the cuttings 15 cm apart along one side of the trench and simply press the soil back against them with the side of the foot. Single cuttings are easily planted with a dibber.

After severe winter weather it may be found that the frost has lifted or loosened the cuttings in the ground, and any affected in this way should be firmed back into position with the foot. Plants rooted this way are generally ready for planting in their final positions in the following autumn.

TIP CUTTINGS

This is a general term, covering all those cuttings which are taken from the growing tip of a non-flowering shoot. They should be treated in the same way as softwood cuttings.

INTERNODAL AND NODAL CUTTINGS

The node of a plant stem is that part of the plant where the leafstalk or petiole joins with the stem. Most cuttings root best if they are cut across in a straight line immediately below a node, and are therefore termed 'nodal' cuttings. This benefit to rooting is especially noticeable in plants inclined to have a hollow stem, as the node generally leads to a branch of solid tissue throughout such stems.

Certain plants though appear to root better if the stem is cut midway between a pair of nodes. These cuttings are therefore termed 'internodal'. Clematis is one plant propagated from this type of cutting (Fig. 6).

IRISHMAN'S CUTTINGS

An Irishman's cutting is a basal cutting which is pulled from the crown of a parent plant complete with a few roots already attached. It is perhaps really a mild form of division.

Early-flowering chrysanthemums and other herbaceous perennials can quite often be propagated in this way. The rooted cutting is potted on with little need to spend any time in the propagating frame, although I would recommend a brief stay until the root system is well established.

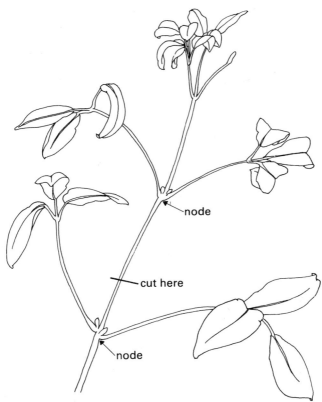

FIG. 6   Internodal clematis cutting.

LEAF CUTTINGS

A wide range of greenhouse plants can be propagated by means of various types of leaf cutting. Usually, a complete leaf, well developed and with its stalk attached, is removed from the parent plant. The leaf stalk is then pushed into a compost of sand and peat until the leaf itself is lying flat against the surface. In a short time roots are developed at the base of the leaf and a new plant is formed. Gloxinias and saintpaulias are two of several plants propagated in this way.

A slightly different treatment is given to the leaves of *Begonia rex*. A crisp mature leaf is taken from the parent plant with

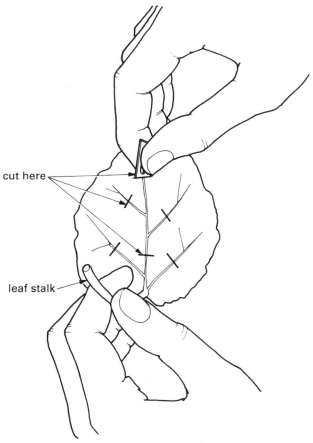

FIG. 7 Cutting the main leaf veins of the underside of a leaf cutting of *Begonia rex*.

merely a stub of leaf-stalk, and a number of cuts 2.5 cm or so apart are made in the well-defined main veins on the back of the leaf (Fig. 7).

With these cuts made, the leaf is pegged down on a peaty compost so that the cuts come into close contact with the medium. Eventually roots will form at each cut and from them new plants will develop (Fig. 8). Such leaves can also be cut into about 2.5 cm pieces, pegging each one to the compost in the way

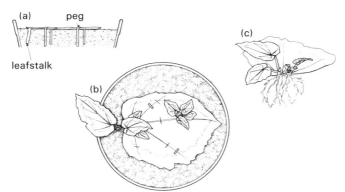

FIG. 8 *Begonia rex* leaf cutting. (a) Peg the leaf down so that the severed veins are in close contact with the compost. (b) New plants form on the cut veins. (c) A young plant separated and ready for potting on.

described. In this way many new individuals can be raised from a single leaf.

Sansevierias (Mother-in-law's Tongue) are propagated in a similar fashion, only this time the leaves are cut across to form sections about 5 cm long. These are planted upright in a sandy compost in the same way as ordinary softwood cuttings.

Leaf cuttings need to be rooted in a close, damp atmosphere; so it is important to keep everything used in the propagating procedure as clean as possible, compost and containers being sterilised if necessary. It is also important to achieve the right balance with watering: too much and the leaves will rot, too little and they will shrivel up. As a general guide to the dampness required take a handful of peat compost and squeeze it tightly in the hand. It should be possible to wring a few drops of water from it if the moisture content is correct.

Most subjects will require a temperature of between 16°C and 18°C to root successfully, so making use of a heated propagator or similar a virtual necessity. The atmosphere in the propagator should be kept damp by a daily syringe with warm water

PIPING CUTTINGS

A piping cutting is simply a stem cutting taken from the tip of a carnation or pink. It is very easy to take, the tip of any non-flowering shoot being suitable.

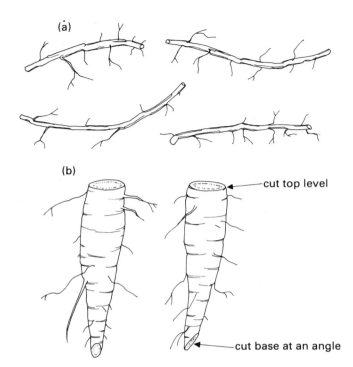

FIG. 9 Root cuttings. (a) Perennial *Phlox*. Such thin types of root are laid horizontally in the compost to grow. (b) Thicker roots such as horseradish should be planted vertically with their tops level with the compost surface.

Hold the selected stem with one hand while pulling the tip carefully with the other, taking care that the cutting parts from its parent just above a node. No other treatment is then necessary, the piping simply being dipped into a rooting compound and dibbled into a sandy compost to root. A cold frame is usually sufficient protection for these cuttings.

ROOT CUTTINGS

Perhaps rather surprisingly, a large number of plants can be propagated from cuttings taken from their roots (Fig. 9), ranging from the Californian tree poppy (*Romneya coulteri*) to the humble horseradish. The method used for all is much the same.

Cuttings are generally taken during the winter months when the plants are dormant. Sections of thicker roots are cut into pieces 5-7.5 cm long, the cut nearest the crown of the plant being made straight across and the lower cut being on a slant, so that you will know which way to insert them. The cuttings are inserted into a suitable compost, such as J.I. Seed Compost, leaving the top and straight end of the cutting just below the surface. A cold frame or greenhouse is all that is necessary for the protection and propagation of the hardy species. By spring the cutting will have branched out and buds will appear at the crown.

A different technique has to be used for thin-rooted subjects, such as phlox. With this type of root it is best to cut them into sections as before but lay them flat on the surface of a box partly filled with a suitable compost, covering them with a shallow layer of the same material. As with the thicker roots, the new shoots will appear in the spring. Plants produced in this way are transplanted into nursery beds in the garden in the following summer.

Certain plants, those of the genus *Acanthus* for example, have a juvenile type of foliage during their early stages of growth. Root cuttings taken from such plants, however, tend to take on the maturity of their parent; therefore, cuttings from a mature plant of this type will develop mature foliage whereas those from a young plant will develop juvenile foliage. As a degree of maturity is required before flowering can take place, it follows that root cuttings from a mature plant are to be preferred.

Another thing to note is that root cuttings from variegated hybrid plants often revert to the green of the original plant, and so should be avoided.

SUITABLE PLANTS FOR THIS METHOD

Season	Plant	Type of cutting	Special requirements
SPRING	*Beleperone* (Shrimp Plant)	tip	propagator
	Cistus (Rock Rose)	basal heeled	cold frame
	Cyperus (Umbrella Plant)	leaf	in water on windowsill
	Dahlia	basal	propagator

SPRING — *continued*

Delphinium (H.P. varieties)	basal heeled	cold frame
Dracaena (Dragon Plant)	tip or root	propagator
Euphorbia (Poinsettia)	tip	propagator
Fatshedera	tip	propagator
Pilea (Aluminium Plant)	softwood	propagator
Sage	half-ripe heel	cold frame

SUMMER	*Achimenes* (Hot Water Plants)	leaf	propagator
	Azalea (*A. indica*=Japanese Azalea)	half-ripe	propagator
	Bay (*Laurus nobilis*)	half-ripe	cold frame
	Begonia (*rex*)	leaf	propagator
	Cacti and Succulents (with segmented stems)	softwood	propagator
	Clematis	half-ripe internodal	cold frame
	Cotoneaster	half-ripe	propagator
	Cytisis (Broom)	half-ripe heel	propagator
	Forsythia	half-ripe	propagator
	Heathers	half-ripe	cold frame
	Hedera (Ivy)	softwood or half-ripe	cold frame
	Hydrangea	half-ripe	cold frame
	Ilex (Holly)	half-ripe	propagator
	Lavender	half-ripe	cold frame
	Lithospermum ('Heavenly Blue')	heel	propagator
	Mint	softwood	outdoors
	Pelargonium (Geranium)	softwood	propagator
	Phlox (H.P. Types)	root	cold frame
	Ramonda	leaf	propagator
	Rhoicissus (Grape Ivy)	half-ripe	propagator
	Rosemary	half-ripe heel	propagator or cold frame
	Saintpaulia (African Violet)	leaf	propagator
	Sinningia (Gloxinia)	leaf	propagator
	Streptocarpus (Cape Primrose)	leaf	propagator
	Weigela	half-ripe	cold frame

AUTUMN	*Acanthus* (Bear's breeches)	root	cold frame (through winter)
	Aubretia (Rock Cress)	softwood	cold frame
	Berberis (Barberry)	half-ripe heel	propagator
	Blackcurrant	hardwood	outdoors
	Buddleia (Butterfly bush)	hardwood	cold frame
	Carnation	softwood	propagator
	Chamaecyparis	half-ripe	propagator
	Cupressocyparis	half-ripe	propagator
	Cupressus	half-ripe	propagator
	Gooseberries	hardwood	outdoors
	Juniperus	half-ripe	cold frame
	Lavatera (Tree Mallow)	half-ripe	propagator
	Roses (Floribundas and Miniatures)	half-ripe	cold frame
	(Climbers and Ramblers)	hardwood	cold frame

WINTER	Chrysanthemums	basal	propagator (through spring)
	Grapes	eye	propagator
	Horseradish	root	cold frame

2 SEEDS

Without doubt there are more plants raised for our gardens from seed than by any other method. Annuals and vegetables are almost exclusively propagated in this way.

There are three main advantages in raising plants from seed. Firstly, seed is a relatively cheap way of obtaining large numbers of plants. Secondly, good seed has less chance of carrying disease than vegetatively-produced plants, and, thirdly, plants raised in the environment in which they will mature will be stronger and more tolerant than their imported cousins.

All seeds, no matter what their type, require the same basic elements to effect their germination. These are moisture, air, light and heat. It is on the degree and manner in which these are given that success or failure depends.

Vegetables and the hardy annuals probably represent the easiest of all seeds to raise. Both types have been bred or selected to give the best results under variable conditions. As long as the basic rules are followed they have a wide degree of tolerance.

It follows that the more exotic the species the more exacting will be its requirements, and in all aspects of cultivation it helps to understand the plant and the environment it inhabits in its natural state. Obviously, the alpine from the high ranges of the Himalayas requires a different set of conditions for germination than a tender plant from the steaming jungles of Brazil.

TYPES OF SEEDS AND METHODS OF ENCOURAGING GERMINATION

Seeds come in all shapes and sizes, from the minute, almost microscopic, seeds of the orchids to the massive 18 kg seeds of the double coconut; but most types fall into recognisable groups bearing the same characteristics. There are five main natural groups: dust-like seeds, hard-coated seeds, fleshy seeds, oily seeds and winged or plumed seeds. There are also the 'pelleted' seeds (Fig. 10).

Pelleted seeds are usually very small seeds which are coated in an inert substance to make them larger and/or a more even shape. This enables you to sow them more easily, either by machine or hand. The seeds can be set out at the correct planting distance from the start. As a general rule, pelleted seed should be sown less

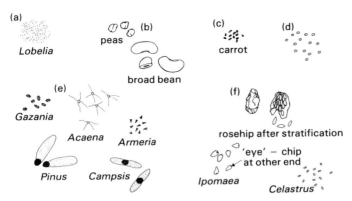

FIG. 10 Different types of seed: (a) dustlike; (b) fleshy; (c) oily; (d) pelleted; (e) winged and plumed; (f) hard-coated.

deep than would be normal for a given seed and at all times the soil or compost used should be kept in an evenly moist condition until germination takes place.

The various types of seeds have differing properties and requirements. *Dustlike seeds*, such as those of begonias, generally do not have a long storage life. They are invariably sown on the surface of the compost and are merely pressed in.

Because their seeds are impervious to water, *hard-coated types* are often a problem as they will not germinate until this protective covering has been broken down. This is achieved in various ways before sowing.

Sweet pea varieties prone to this trait are often 'chipped', which means that a small sliver of the skin is removed from the seed opposite the eye. Care should be taken not to damage the seed with too enthusiastic a cut, and it is perhaps safer to use a small file to wear away a portion of the surface.

Freshly gathered hard-coated and flesh-covered seed such as the holly produce better results if they are 'stratified'. This is done by placing the seeds in layers in a pot or box containing a mixture of sand and peat in equal quantities. These are then placed outside where frost and snow can work on the seed coat through the winter and soften it. Mice and birds are the enemy here and containers are best protected with a covering of fine wire netting.

Many alpine seeds require this type of cold treatment. Normally it can be achieved by sowing early in the year and exposing the pots to weather throughout February and March. This can be artificially reproduced by placing the seeds, either in their moistened seed packet of preferably already sown in a moist compost, leaving them overnight to absorb the moisture before placing them in the freezer compartment of a fridge for a time. Forty-eight hours is said to be enough, but I always leave mine in for around a week. It is most surprisingly effective.

Quite the opposite of cold is the treatment required by some eucalypts, brooms and other heath/forest plants which only germinate freely after being subjected to fire and heat in some degree. Such seeds can be given this treatment by placing them in a metal sieve and passing them over a gentle flame until a few burst, indicating that they have reached the required temperature. When this happens the remainder are plunged immediately into a container of slightly warm water for a few moments before sowing in the normal way.

Fleshy seed is the term used to describe such seeds as peas and broad beans. These respond well to being soaked in clean water for about 24 hours before sowing. Incidentally, this method will also determine whether a variety of sweet pea is hard-coated or not — those seeds which require chipping remain afloat after their more receptive cousins have absorbed moisture and sunk.

There is little difficulty with the germination of *oily seeds* other than to note that such seeds soon shrivel. Carrot and magnolia are of this type and should be sown as soon as they are ripe, or the season they are received as they will not keep well in storage.

Lastly come those seeds intended for dispersal by the wind. Such seeds have wings as in those of the sycamore, or plumes as in seeds of the dandelion. With *winged seeds* the wings are best removed before sowing.

Temperatures for optimum seed germination are as varied as the seeds themselves, but generally most respond well to a temperature of 5-6°C higher than the natural temperature in which the plant achieves full growth. In general *very* high temperatures are to be avoided unless specified as they can inhibit germination rather than encourage it, particularly with certain plants, such as lettuces.

Although the seed of highly bred garden varieties is likely to germinate relatively evenly and quickly, the seed of many species is often slow to germinate. Some seed takes two years or more, and with such types it is best not to assume failure until at least this amount of time had passed. During this waiting period it is essential to keep the seed compost moist and to protect the seeds from the attack of mice, birds and insects.

COLLECTING AND PREPARING SEED

Whereas our climate allows us to grow numerous plants to their flowering stage, in many cases it is a rare year when we are graced with a summer and autumn warm and dry enough for their seeds to ripen adequately. This and various other factors make seed-saving by the amateur a rather restricted affair.

Nevertheless, to grow a crop from your own seed gives a great deal of satisfaction and many plants do oblige by setting seed, some more enthusiastically than would be wished.

As seed ripening follows flowering it will be obvious that those flowers and vegetables reaching maturity earlier in the year will be the most likely to set seed and ripen. In the vegetable garden peas, various kinds of beans, lettuces and tomatoes all oblige by maturing early on self-pollinating plants. Biennial vegetables, too, such as beetroot, carrots and onions, are also suitable, being grown on to flowering in their second season after winter storage.

One of the difficulties in the average garden is making sure that plants are not cross-pollinated, which is why those that are self-pollinating are best suited. The cabbage family in particular cross-pollinates easily, and seed saved from any could yield a hotch-potch of cabbage types, mostly useless, and none or very few showing the characteristics of its immediate seed parent. If an attempt is made with these types of plant it is essential that only one variety should be allowed to flower, thus giving the likeliest chance of true seed (and even then F_1 hybrids will not produce offspring like the parent).

Plants selected for seed production should be chosen for their health, vitality and closeness to the 'type' of plant required. With tomatoes, for instance, you could select the plant bearing the tastiest fruit or conversely one bearing the heaviest crop. In an ideal world the two might be combined. When selecting always remember that characteristics given to a plant by its growing environment will not be passed on to its seed.

Peas and beans are perhaps the easiest seed for the beginner to save as they are easily seen and handled. Seeds should be saved from those pods appearing first, thus giving more time for successful ripening. A part of a row can be set aside for the purpose or the lower pods left to ripen while those above are picked.

With all seed-saving ventures it is important to mark the plant or plants selected as seed parents clearly. The 'head gardener' is not always the one who picks the vegetables or flowers and many a prospective seed harvest has ended up in the pot or the table decoration!

Dry weather is essential for good ripening and in the case of peas and beans these are known to be ripe when the pods begin to look dryish and yellow and begin to split (Fig. 11). In wet seasons the plants are best pulled up whole just prior to ripening and hung up in a well ventilated shed or greenhouse where the rain cannot reach them.

When dry the pods are taken singly from the plant as they split and the seed removed, placing them in shallow containers to dry fully in a warm greenhouse, shed or on a sunny windowsill. As with most seeds, when dry they should be stored in unsealed paper bags in a cool, dry and airy place. At all times seed stocks should remain clearly labelled.

The seeds of tomatoes, and of other similar fleshy fruits, are more complicated to prepare for storage, the first essential being that the fruit should be really ripe before it is picked. Tomatoes are very soft, almost mushy to the touch when they are fully ripe.

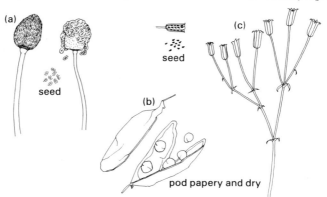

FIG. 11 Ripe seedheads. (a) *Anemone coronaria*, unripe (left) and ripe (right). (b) Ripe peapod. (c) *Aquilegia*.

Vegetable marrows on the other hand are hard and sound slightly hollow if lightly tapped with the knuckles. It is difficult to know whether cucumbers are ripe or not, and it is best to leave them hanging on the plant until well into autumn before preparing them.

Commercially, these types of seed are removed from the fruit, first by pulping them (not crushing them!) in a wooden or earthenware container and then treating them with commercial hydrochloric acid at the rate of 1 fluid ounce to 500 g of fruit. This mixture is left for varying periods depending on the plant and separates the seeds from the pulp after which the seeds are washed thoroughly in fresh water.

This method can be used by anyone, of course, but a less costly and safer method is best suited for home use. The ripe fruit is pulped in the same manner into a bottling jar or similar and allowed to ferment in a warmish place for two or three days, with an occasional stir. This will cause the seed to separate from the pulp to a large degree whereupon the whole mess can be placed in a fine sieve and the seed washed clean under running water. After separation and washing, the seed is spread thinly on a clean cloth to dry in a warm place before storing.

Seeds can be saved from the many ornamental garden plants which are true species. Seed from hybrid forms and varieties should be avoided as they are unlikely to come true to type. Such plants are best propagated vegetatively from cuttings etc. Having said this, however, it must be remembered that most hybrids have been raised from seed originally, and experimenting can be fun if you have the time and space.

Nevertheless, if we leave aside the various hybrid types of plants in our gardens, a quick inventory will reveal many which are true species, and amongst these are many shrubs, trees, alpines and border plants.

One of the advantages of collecting your own seed is that it can be sown as soon as it is ripe, often resulting in much improved germination. Primulas, for example, are notorious for slow and haphazard germination after storage, yet fresh seed germinates easily. With all self-saved seed of species plants it is a good plan to sow a portion of it as soon as it is collected and store the remainder for sowing in the spring.

Keen observation and dry weather are the two real necessities

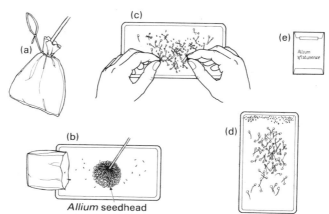

Allium seedhead

FIG. 12 Preparing seed for storage. (a) The seedhead tied inverted in a paper bag for final drying, and hung in a warm spot. (b) When dry, empty the contents into a small metal or plastic tray. (c) Roll the seedpods between the fingers to release the seed. (d) Shake the tray gently from side to side to separate the seed from the chaff. (e) Label clearly and store in a cool, dry place in unsealed paper envelopes.

for successful seed collection. A careful watch should be kept on the ripening seed heads so that they can be plucked from the plant just before the seed is shed. A large number of plants shed their seeds explosively, the pod or fruitcase bursting to fling the seeds everywhere. These types are best collected a few days early and the cycle completed in a controlled way by inverting the seed heads in a paper bag suspended in a dry, warm place in the sun (Fig. 12).

Separating seed from chaff is fairly simple in most cases. When perfectly dry both seed and chaff should be placed in a shallow tray and gently rolled from side to side, at the same time blowing over them in the same gentle manner. Any unbroken seed cases can be rubbed between the fingers to extract the seed. Large pieces of chaff and other waste are best removed with the fingers or tweezers as the process continues.

Seed contained within fleshy fruits such as rosehips should not be dried out, as germination can prove difficult when the seed is eventually sown. Such seeds are best sown as soon as they are collected, or stratified as described earlier.

In the spring the sand/peat mixture in which the seed has been placed should be turned over every few days with the fingers. When a few seeds are seen to be sprouting the whole mixture should be sown in an appropriate manner for the particular species.

Bulbous plants very often set seed freely, and those species flowering early in the year are amongst the easiest of all to collect. Unfortunately, it does require patience to see the results of your labours as they often take several years to reach flowering size. To offset this the enthusiast should collect and sow some seed every year, then after the initial few years of waiting subsequent seasons bring forth their results annually. Once more bear in mind that those plants bearing varietal names, such as 'King Alfred' daffodils, are hybrids and will not grow true from seeds.

F1 HYBRIDS AND HYBRIDISATION

The many hybrid plants in our gardens are generally the result of crossing one species of plant with another in the same genus, or between distinct varieties within a species. An example is the polyanthus, which originated as a cross between *Primula acaulis* and *Primula veris* (the primrose and the cowslip, the wild hybrid being known as the false oxlip).

Almost all modern strains and varieties of such plants are the result of selecting and/or crossing the seedlings which resulted from the original cross.

Hybridisation is a complex subject and the results from many crosses are unpredictable, which is why a great many seedlings have to be raised to find that one plant with all the virtues required. This, as well as the necessary trials and selections that follow, accounts for much of the cost of 'novelty' introductions.

The first seedlings to be raised from a cross between two plants are called the first generation, or F1. Subsequent generations of crosses from these F1 plants are labelled F2, F3 and so on.

Although F1 is a term used for all first generation seedlings it is particularly significant when attached to the F1 hybrids increasingly offered in seed catalogues. These hybrids are the result of crossing two parents with distinctive qualities, the first generation bringing out the best of both. The parents used for this type of seed production are plants which have been grown and selected over many generations until they are a pure line, this meaning that they always breed true from seed and produce virtually identical plants.

This process generally leads to a weakening of the plant characteristics, but, rather surprisingly perhaps, when two distinct pure lines are crossed they yield a first generation of extremely strong, often disease-resistant offspring which show very little variation between individuals.

Great skill is needed to raise F1 hybrids, as both parents often have to be isolated to prevent accidental cross-pollination, and in many cases pollination has to be done by hand under artificial conditions. This, and the care taken in selecting and maintaining the distinct pure lines which provide the parents, is once more reflected in the cost of such seed. Nevertheless, anyone growing plants from F1 hybrid seed can hardly question the value received in results.

Seed saved from F1 hybrids will provide a mixture of all the genetic forms within the plants, many of which will be completely useless. For this reason seed is not worth saving.

Occasionally, among such plants as chrysanthemums, plants appear either wholly or partly changed in colour from their varietal norm. Such plants are mutations or 'sports' which appear for no apparent reason. These will not breed true from seed but promising variations can be propagated vegetatively and many new varieties have been raised in this way. They are not to be confused with 'stray' plants of another variety.

SOWING SEED

Seed is sown in two basic media; either in a specially formulated compost for growing under glass or in the ordinary soil of the open garden.

In many respects sowing in compost under glass has more advantages, the main one being that sowing is less dependent upon weather conditions and plants started into life early in the year generally yield a better crop. With many plants it is essential that they are started into growth under glass, indeed, many are best grown in such conditions throughout their lives.

In all cases under glass the first essential for successful seed raising is cleanliness, as the main hazard facing seedlings in a warm, relatively still atmosphere is an attack by one or more of the various fungus diseases. The main offenders are those causing

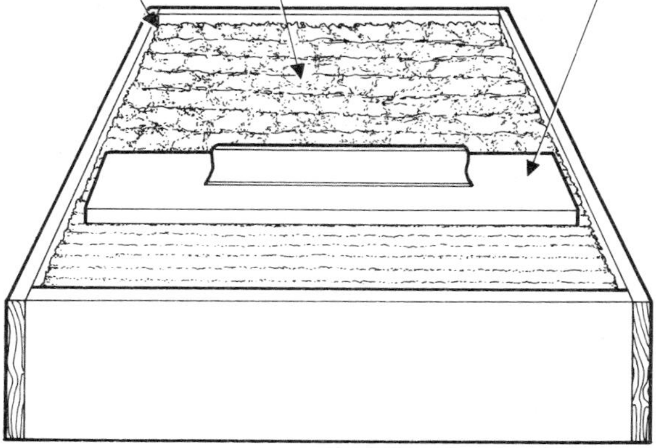

FIG. 13 Preparation of seedbox prior to sowing or pricking out small seedlings.

press down corners with fingertips

level roughly

finally, firm gently with a piece of wood

the condition known as 'damping off', symptoms of which are rotting-off of the seedling at ground level and its subsequent collapse.

The risk is lessened by using sterile containers and compost, plastic being preferable for the former as it is easier to clean than other materials. Should soil be used in the compost then this too should be sterilised. A simple way of sterilising a small quantity is to fill a large saucepan with dust-dry soil, putting it in on top of 1 cm of water. Bring this to the boil, then simmer for fifteen minutes. Turn the soil out on a clean place to dry until it can be handled, before mixing into compost. For larger amounts various sizes and types of commercial sterilising units can be obtained or for the greenhouse and outdoor beds various chemicals can be used to good effect.

Strong healthy seedlings are less prone to fungus diseases, and such plants can be encouraged by sowing thinly and keeping the seed compost evenly moist without allowing it to become waterlogged. A preventive spray with either Cheshunt Compound or a liquid copper fungicide immediately prior to sowing, and again when pricking out, can be nothing but beneficial in the majority of cases.

It is important to use a good seed compost and there are many proprietary brands on the market or the more enterprising can easily make up his or her own. The John Innes formula for seed compost (*see* page 9) or soilless compost suggested for cuttings should prove adequate for the most part, but the plants must be pricked off without delay once they germinate.

Before use all compost should be evenly moist but not saturated unless culturally necessary.

Before making a start it is best to prepare properly by making sure that all materials are at hand and labels are written. Fill the pots or boxes with your chosen compost, filling them almost to the brim, and press down the corners and sides in the case of boxes to dispel any air pockets, settling the contents of pots with a couple of light taps on the workbench top. Finally, level the compost with the fingers and firm it lightly with a wooden 'patter' or the base of a clean pot (Fig. 13).

Sow the seed on this surface as evenly as possible, ideally by placing each seed individually with the fingers with the larger types or by sprinkling them from the hand between thumb and forefinger. Smaller seeds are best sown direct from the packet, the smallest being mixed with a little clean dry silver sand to make them more visible. It aids more accurate distribution if the tip of the packet is cut across cleanly with scissors, and one side bent to a V-shape with the fingers. This forms a channel through which the seed can be shaken in a controlled way.

Very small dust-like seeds should be merely firmed into the surface of the compost with a patter, whereas large seeds need to be covered in compost to various depths. A good general guide is to cover the seeds with the same depth of compost as their girth or thereabouts, but unless otherwise specified never sow seed too deeply or it may fail to germinate altogether (Fig. 14).

Some seed companies offer pre-sown seed — all you need to do is add water and wait. The more recent ones use a specially prepared vermiculite in a plastic container. If you find sowing a chore they are worth considering, but you pay more for the seeds of course, and the range available is small. You still need to provide sufficient warmth and light.

If it is the pricking out rather than the sowing that you find

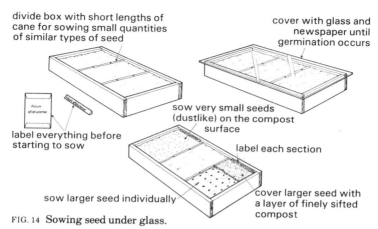

FIG. 14 Sowing seed under glass.

difficult, you can even buy seeds attached to a piece of card that you push into the compost at the right spacing (there is a depth mark on the 'stick' too). There may be several seeds on a 'stick' but you simply thin to one if more germinate. The drawbacks are the same as those for pre-sown seed, already described.

Compost used to cover the seed should be sifted over them carefully with the fingers taking care not to use too fine a compost as this is inclined to cake and prevent air reaching the seeds.

With the seeds sown, the pans and/or boxes are grouped together and given a gentle watering with a fine spray before being covered with sheets of glass, to conserve moisture, and in many cases with paper, to exclude very bright light. Newspaper is perfectly adequate for this.

In many cases no further watering will be required until after the seeds have germinated, and a daily watch is essential to note when this occurs. As soon as a few seedlings have appeared both glass and paper are removed taking care that the now unprotected seedlings are in good light but shaded from full sun.

Never leave seedlings covered with glass or paper once germination has begun or they will become drawn and leggy within a very short space of time. Such plants never make satisfactory mature specimens.

Once the seedlings are established they will need transplanting or 'pricking out'. The sooner this is done the better but I leave it until the first true leaves begin to appear when the seedlings can be handled with comparative ease.

Very small seedlings are best pricked out into the same type of compost as they were germinated in, feeding them as necessary as they become established. Stronger growing types need a stronger compost, J.I. Potting Compost No. 1 or its equivalent being ideal. The J.I. potting composts are based on the following formulae:

7 parts (by volume) medium loam (sterilised)
3 parts moss peat
2 parts horticultural grade coarse sand

For J.I. No. 1 add the following per bushel:

110 g of fertiliser made up of:
2 parts (by weight) hoof and horn meal
2 parts superphosphate of lime
1 part sulphate of potash
plus 20 g ground limestone

As with the composts, the fertiliser can be bought made as John Innes Base Fertiliser. As a guide to compare with other suitable proprietary brands its approximate analysis is: 5% Nitrogen, 7% Phosphoric Acid and 10% Potash.

The same procedure as for preparing the seed containers is followed for pricking out, filling the pots or boxes almost to the rim and firming them to exclude any air pockets. The compost used should be evenly moist and warmed to the greenhouse temperature, this being especially important in the early part of the year. It is easily done by preparing the compost the day before using it and leaving it on the greenhouse bench overnight.

Seedlings should be pricked out 2.5-7.5 cm apart depending on their strength of growth and size. Avoid disturbing the roots too much by teasing them out of the compost with a pointed stick. Always hold the seedling by the leaves and never the stem which is all too easily bruised. Small seedlings can be supported using a thin slice of wood with a V notch cut in its end, called a 'widger' (Fig. 15).

A hole is made to receive the roots with the same pointed stick as was used to lift them, taking care that the roots are not doubled back on themselves. Seedlings with exceptionally long

small seedlings can be lifted using a small notched sliver of wood called a 'widger'

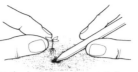
make hole large enough for roots with a pointed stick

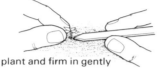
plant and firm in gently

FIG. 15 Pricking out seedlings.

roots can be made much easier to handle by dipping them in a small pot of mud made from sterilised soil and water. By giving them a twizzle as they are removed, the roots cling together whereupon they are easily twirled into the hole made to receive them. This technique is also useful outdoors for such plants as onions.

After pricking out, the seedlings should be kept warm, moist and lightly shaded until they are established, spraying them over daily with a fine spray if the weather is hot. If the original compost was sufficiently moist they should not need a real watering for a few days.

Seedlings which are destined to be planted outdoors to mature are grown on until the weather and season is right for them to go outside, hardening them off to outdoor conditions by gradually cooling them. This is best done by transferring them to a cold frame several weeks before planting out, gradually opening the frame's lights for longer periods until they are taken off altogether. Care is needed at this stage to prevent the plants being either baked in the hot sun or frosted by a sudden spring frost. The former can be prevented by shading, the latter by covering the cold frame lights with sacking or old carpet.

Plants to be grown in pots for either greenhouse culture or the home are potted on when they have outgrown their pricking out positions. Pots of 7.5 cm are usually used at first with John Innes Potting Compost No. 1 forming the growing medium.

As the season progresses such plants are repotted into stronger mixtures and larger pots, John Innes Potting Composts Nos. 2 and 3 being made to the same basic formula as No. 1 with only the fertiliser and lime content increased. With No. 2 this is doubled and with No. 3 trebled.

SOWING OUTDOORS

The main factor governing outdoor sowings is the weather. It is no use going ahead with sowing if the ground is cold and wet, far better to wait until there is a definite improvement and the soil warm and workable. There is nothing to be lost by waiting a while and much to be gained in better germination and stronger growth.

Seed can be sown in the open ground by either broadcasting it or sowing it in drills.

The broadcast method is usually reserved for lawns and the odd bare patches in the garden where it might be advantageous to sow a few annuals, perhaps between newly-planted shrubs which will take a few years to establish themselves.

As its title suggests the method is simply executed by sprinkling the seed in an even but haphazard manner on the soil surface, the process being completed with a light raking in opposite directions to settle the seed in a little. Birds can be a problem and are best kept at bay by stringing black cotton about the area. Some lawn seed is also treated with a bird repellent.

Sowing in drills is a more exact science, necessitating in the first instance a string line to make sure that the drills remain straight and parallel. If there is a place for neatness and regimentation in the garden then it is the vegetable and nursery plot.

Drills are made at variable distances apart and to various depths depending on the type of plant to be grown. It is usual to sow outdoor seed slightly deeper than those under glass.

Drills are easily made with the corner of a hoe or rake, drawing the tool slowly along by the side of the line, being careful not to move it out of position. A few extra canes placed along the line helps to avoid this on longer rows.

In some cases, such as pea rows, the drill becomes extended into a bed some 15 cm wide by 5 cm deep. This is best made with a spade, being careful to keep to the same depth throughout the length, as with the smaller drills.

Before starting any outdoor sowing programme the soil should be well prepared and broken down to a fine tilth with a rake or cultivator. If artificial fertilisers are being used then these should be worked into the soil a few days before sowing is to take place. If lime is also being added the two must never be put on together as they react unfavourably when in close contact. Incorporate one into the soil before adding the other.

This done and the drills made, the seed is sown thinly and carefully in the same way as sowing under glass. Again large seed such as peas and beans are best placed individually.

Once sown the seed is covered by gently drawing the soil back

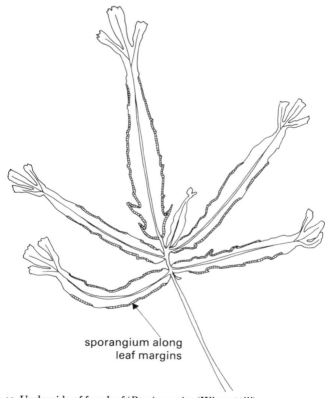

FIG. 16 Underside of fern leaf (*Pteris cretica* 'Wimsettii').

over them with a rake, and I always like to follow this by lightly firming the row with the back of the rake before moving on. With the line at its next station a row so marked is easily seen. Always leave a cane at each end of all rows as a marker, with a clearly written label. Memory is a most doubtful ally for the gardener.

Should the soil be dry then it is best to water the drills before sowing, preferably some time before to allow the moisture to spread. Never water the seeds in as this may cause panning of the soil surface which will effectively seal out the air necessary for good germination.

Once the seedlings are up and growing strongly they will need thinning. This is done by removing those plants which are not wanted, the final distance between the plants being determined by their type.

Thinning is best done in stages, the first when the plants are very small. This consists of taking out any obviously weak or unnatural specimens as well as reducing the numbers in those 'clumps' which always seem to appear in hand-sown rows. A week later a few more can be removed, again taking out the weaker plants but also bearing in mind the eventual spacing. Some time later a final thinning ideally leaves a row of evenly spaced individual plants.

Seed can also be sown in drills using a mechanical seed sower which is most helpful if large quantities are to be sown. They are generally fitted with an adjustable spacing guide which makes the line redundant once the first drill has been made.

SPORES

Spores are the reproductive units of the fungi, mosses, lichens and ferns which perform a similar function to that of seeds in the higher plants.

The spores of ferns are very small indeed, virtually microscopic, and are produced in vast quantities by many fern species. They are produced on the underside of the fern leaves or fronds, sometimes on special leaves adapted for spore production (Fig. 16). The clusters of sporangia are called 'sori'.

Most spores are easily collected by the amateur, the same principles as in seed-saving being applied. As spores are so light and easily dispersed when ripe it is important to keep a frequent watch on them as they ripen, this being indicated by a gradual browning of the sporangium which bears them, though in a few genera (such as *Osmunda* and *Matteuccia*), the spores remain green even when ripe.

These latter, however, are less common than the brown, and as their spores are generally short-lived, propagation with them is best left until some experience has been gained. With the more general types, the spore-bearing fronds should be cut from the plants as soon as the browning is pronounced, placing them in a clean paper bag for their final development, and afterwards fastening the top securely to prevent the spores escaping into the air.

Best results can be expected from freshly ripened spores and they can be sown at any time, although spring is probably the best. Greenhouse ferns will require a temperature of 18-21 °C for germination, whereas hardy ferns are happy with a little less.

As the germination of ferns is biologically complicated it can take some months for them to reach transplanting size. As the conditions favourable for fern spore germination are also favourable to other unwanted spores, everything used should be completely sterile. Once in sterile pots the compost should be treated by pouring boiling water over it, leaving it a few minutes to cool before sowing the spores.

A suitable compost is one consisting of equal parts by volume sterilised loam, moss peat, leaf-mould and sand; or the John Innes Seed Compost mixed with a further two-thirds (by bulk) moss peat.

Pans of 13 cm make suitable containers and should be filled almost to the top with compost, levelled and firmed and then treated with the boiling water. When cool, sow the spores thinly and evenly over the surface of the compost (a job best done in good light and away from draughts) and press them down slightly into the surface with the base of a sterile pot. When finished the pot should be stood in a dish of water to maintain the compost in a moist condition, placed in a shady position, and covered with a sheet of glass, this being removed daily to wipe away the condensation to avoid drips.

After a number of weeks a green slime should appear on the compost surface. This is the first stage of the fern's growth. The slime is made up of large number of separate units called prothalli. These continue to expand for some time, and when their growth appears to have ceased pricking out can begin.

This is done by carefully removing 5 mm sections of the prothalli and transferring them to other pots prepared in the same manner as before. Space the pieces out on the surface of the compost about 2.5 cm apart and simply work them in, making sure that they are never planted any deeper than they were originally. The pots are then covered with glass and stood in water as before.

After about a week start to ventilate the pots by raising the glass slightly, giving them more air each day until by the time the new ferns can be seen they are ready to be left in the open air entirely.

When large enough to handle the young ferns are potted individually into small pots, potting on as soon as the pot is filled with roots, as most ferns resent being starved in their early stages.

SPAWN

Though mushrooms belong to the fungi group of plants and so reproduce naturally by spores, a vegetative method of propagation is used by both amateur and professional growers.

Mushrooms themselves form only a small part of the entire plant, being only the fruiting part to bear the spores. The main life of the plant is carried on below ground as a mass of white threads called the mycelium. It is with small sections of this mycelium that the gardener begins, this being supplied either in dry 'manure blocks' or inoculated into inert grains of rye or millet. Both types are known as spawn.

The production of spawn is very specialised indeed and is undertaken in sterile laboratory conditions. For this reason even professional growers obtain their spawn from specialist producers.

Traditionally, horse manure is the basis of mushroom compost, but nowadays there are also artificial methods of producing it, these usually from a straw base. Horse manure is treated by stacking and fermenting it for 10-14 days when the temperature within the heap should rise to around 65-71°C. When this is reached the heap is turned so that the outside becomes the centre, shaking it out and watering with a fine hose if it appears dry. Do not overwater.

The heap is turned twice more at 4 day intervals and at the second turning it is generally recommended that a certain amount of gypsum (calcium sulphate) should be added. The amount varies, but a rate of 1 kg gypsum per 50 kg of compost is reasonable.

After this time the compost is moved to the growing bed, treading it down in two layers to form a complete bed 15-20 cm deep.

The spawn is added when the compost temperature falls to around 21°C, block types in plum-sized lumps 2.5 cm deep and 23 cm apart, grain types spread on the compost surface at a rate of around 305 g to 1 m^2 working it into the compost with the fingers and firming in.

About two weeks later the bed is 'cased' with sterilised soil or moss peat/ground limestone in equal quantities by weight to a depth of 3.8 cm.

Block spawn can also be used to supplement naturally-occurring mushroom colonies in lawns or fields by sowing small pieces just below the surface of the turf in warm, moist weather during summer; results, however, are unpredictable, depending as they do on the weather.

SUITABLE PLANTS FOR THIS METHOD

Season	Plant	Special requirements
EARLY SPRING	*Abies* (Silver Firs)	outdoors
	Alyssum (*Lobularia*)	propagator
	Antirrhinum (Snapdragon)	propagator
	Begonia (fibrous-rooted)	propagator
	Brussels sprouts	outdoors
	Carrots	outdoors
	Celery	propagator
	Cucumber	propagator
	Fern spores	propagator
	Helianthemum (Rock Rose)	cold frame
	Impatiens (Busy Lizzie)	propagator
	Leek	outdoors
	Leontopodium (Edelweiss)	cold frame
	Lobelia	propagator

EARLY SPRING — *continued*

	Melons	propagator
	Parsnip	outdoors
	Primula Obconica	propagator

SPRING	*Ageratum* (Floss Flower)	propagator
	Anemone (Windflower)	cold frame
	Aquilegia (Columbine)	propagator
	Beetroot	outdoors
	Cabbage (summer and winter)	outdoors
	Calendula (Pot Marigold)	outdoors
	Cauliflower	outdoors
	Godetia	outdoors
	Iberis (Candytuft)	outdoors
	Nigella (Love in a Mist)	outdoors
	Onion	outdoors
	Parsley	outdoors
	Petunia	propagator
	Vegetable marrow	propagator

LATE SPRING	*Calceolaria* (Slipper Flower)	propagator
	Cheiranthus (Wallflower)	outdoors
	Courgettes	propagator
	Digitalis (Foxglove)	outdoors
	Lupinus (Lupin)	outdoors
	Primula Malacoides	propagator
	Sweet William	cold frame

SUMMER	*Arabis* (Rock Cress)	cold frame
	Bellis (Daisy)	outdoors
	Cyclamen (Greenhouse varieties)	propagator
	Leaf beet	outdoors
	Myosotis (Forget-me-not)	outdoors
	Spring cabbage	outdoors
	Viola (Violas/Violets/Pansies)	propagator

AUTUMN	Broad bean	outdoors
	Sweet peas	propagator

WINTER	*Androsace* (Rock Jasmine)	stratify outdoors
	Betula (Silver Birch)	stratify outdoors
	Gentiana (Gentian)	stratify outdoors
	Primula (H.P. Types, Primrose etc.)	stratify outdoors
	Tomatoes	propagator

3 DIVISION

The division of plants which are already established in the garden is probably the easiest of all methods of propagation and certainly one of the most successful for the beginner. As suggested by its name, the method involves the dividing up of an existing plant, each part divided being complete with its own roots and growth buds.

Division is most often done during autumn or early spring when the plants are dormant, and is used for the increase of many perennials, alpines, bulbs and houseplants. The great range of plants which can be divided naturally leads to a number of ways of achieving the desired result. It is easiest to divide these methods into six basic types, each dealing with a particular group of plants having the same growth characteristics.

FIBROUS-ROOTED PLANTS

Many perennials in the herbaceous border will be of this type; Michaelmas daisies, anthemis, artemisias and heleniums being just a few of them. After a few years in the border many plants will have overgrown their space and will possibly begin to die back in their centres. For their own good these need to be lifted and divided.

Though most plants can be divided in autumn or spring some are best done in spring only, pyrethrums and scabious being amongst them.

With small, loose-rooting kinds of plant it is a simple matter to separate the roots with the hands, pulling a small clump containing a few crown buds away from the parent plant (Fig. 17). Others may need the assistance of a sharp knife and some more drastic treatment still.

Should fingers or small knives seem inadequate when faced with a large and extremely tough root system, then a division will have to be made with the aid of two forks. Either hand forks or full-sized forks can be used, depending on the strength of the gardener and the resilience of the root system to be tackled. The method used is identical in all situations, the forks being driven through the centre of the clump back to back and then carefully levered apart.

Sometimes, especially with long established plants, it may be

FIG. 17 Dividing the fibrous roots of a polyanthus.

found that their centre has become woody and largely unproductive. In these cases the centres should be discarded, only the fresh new growth around the perimeter of the crown being replanted.

A mention might be made here about weeds. As many perennial weeds such as couch grass are the bane of the herbaceous border, it is often only at such moments as when plants are being divided that the troublesome individuals can be removed with any success and every effort should be made to do so before replanting the new divisions.

WOODY-ROOTED PLANTS

Some herbaceous perennials will be found to have very woody

roots and crowns and may need rather more drastic treatment to effect a division than other types. Established plants of this kind should be lifted in spring before the plant has started into growth in order to carry out the division.

It helps greatly with the task if the crown of the plant is washed thoroughly once lifted. This enables one to see the buds easily so that the division can be made in the most favourable places. This can usually be done with a strong, sharp knife, the plant being cut up into appropriately sized pieces.

Should a knife or your muscles seem inadequate for the job then a spade or curved edging iron can be used. Although treatment like this may seem somewhat harsh the plants will soon make good once planted in their new position. Less damage will occur if the cut is made without any hesitation. Just choose the best place to make the cut and give it one heavy chop. One enormous slicing cut is far better than several bruising half-hearted attempts trying to do the job gently.

OFFSETS

An offset is a young plant produced alongside the parent and easily detached from it. There are several kinds of offsets. Some cacti and succulents, for instance, produce a cluster of small plants around the base, and these offsets merely need separating and potting up. Bromeliads produce offsets around the old flowering shoot, and bulbs produce offsets freely.

Everyone who has had any dealings with bulbs will undoubtedly have seen smaller bulbs or bulblets growing from the base of the larger specimens. Daffodils in particular can be seen in any garden shop with several growing tips projecting from apparently one basic bulb. Such bulbs are called 'mother bulbs', and if the smaller bulbs growing from them are removed and grown on they will become individuals in their own right and eventually produce bulblets of their own. These are known as bulb offsets, a natural method of propagation which the gardener can make use of to good advantage (Fig. 18).

Some plants such as crocus and gladiolus produce corms which, although very similar to bulbs in appearance, are completely different internally. These are formed from a swollen stem, whereas bulbs are formed from modified leaf bases. With corm-bearing plants, new corms or cormlets develop on the top of the

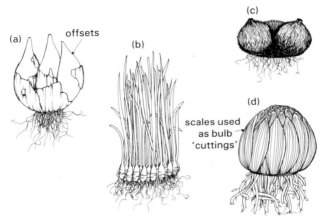

FIG. 18 Offsets forming on bulbs and corms. (a) Tulip. (b) Chives (*Allium schoenoprasum*). (c) *Ixia*: cormlets growing on top of the old corm. (d) *Lilium*.

old corm or around its base and can be removed and grown on the following season as before.

Many bulbs and corms are lifted from the garden or pots during their dormant season and this is a convenient time to remove any offsets which might have appeared.

Liliums form a different kind of bulb again and can be increased by means of their scales which are broken off at transplanting time. It is debatable whether this method is really a form of division and should perhaps come under the heading of bulb 'cuttings', but nevertheless this is the method to use. John Innes Seed Compost is a suitable medium, the containers used not being filled to capacity. Cover the compost with a thin layer of silver sand, then stand the broken-off scales upright like cuttings, 2.5 cm or so apart, before filling over with sand until the tips of the scales just show above the surface. They can be dibbled into the sand with a stick or label if you prefer, but always leave some space in the box or pot for top dressing as the new bulblets form. The pots or boxes should be placed in slight warmth to encourage the formation of the bulblets. Do not overwater.

Many other bulbs can be propagated by means of their internal scales or layers and the enterprising amateur can gain much satisfaction from experimenting. Bulblets can be formed by mixing scales with a mixture of equal parts peat and silver sand, enclosing them in a polythene bag tied at the top, and incubating them in a warm place until bulblets have formed. These are then removed and grown on as for seedlings. Some, such as daffodils, may require a period of cold growing at around 4°C following the incubation period to encourage rooting.

RHIZOMES

Quite a number of plants have underground root-like stems which, by storing food, enable the plant to survive during the winter and to propagate itself by vegetative means. The weed couch grass is one such plant, the underground stems spreading the plant far and wide, as many gardeners know to their cost. Happily, not all such plants are weeds.

Among the cultivated plants the bearded iris is probably the most recognisable of the rhizome-bearing types. The rhizomes of these plants are easily divided by cutting pieces off the underground stem of the parent plant after lifting. The pieces are formed by cutting sections 5-7.5 cm long from the parent plant, making sure that each part has at least one strong growing shoot rising from it (Fig. 19).

With many rhizomes the best portions for propagation come from the outer edges of the plant, the original centre section being discarded. The retained portions are replanted in a horizontal position just below soil level. With most types the operation is carried out after the plants have flowered, usually around July.

SUCKERS

A sucker is the name given to the shoots sometimes arising as a type of offset from the root systems of certain plants and trees such as raspberries and poplar. These suckers often surface some distance from the parent plant and can be mistaken for seedlings.

Although more often than not a nuisance, these suckers can be used as a means of increasing stock, and are simply separated from the parent root and transplanted to a new position. This is generally done in autumn or early spring, the normal planting time for these types of plant (Fig. 20).

The job is usually simple enough, but lift the shoot carefully, cutting the sucker away from the original root with a piece of that root still attached to the new root base.

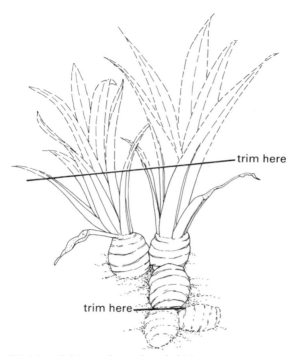

FIG. 19 Division of rhizome from a bearded iris.

Many plants can be increased in this way but it must be remembered that suckers rising from grafted plants such as lilacs and roses will produce plants of the *rootstock*, which is most unlikely to have the desirable characteristics of the grafted portion.

TUBERS

Tubers are formed from swollen underground stems or roots and are used by the plant for storing food, the commonest example of the type being the potato.

Tubers are quite simple to divide but as they come in a number of different forms handling them may, at first sight, seem more difficult than it really is. One essential in all cases is that, wherever the dividing cuts are made in the tuber, all pieces cut away should contain at least one crown bud or shoot.

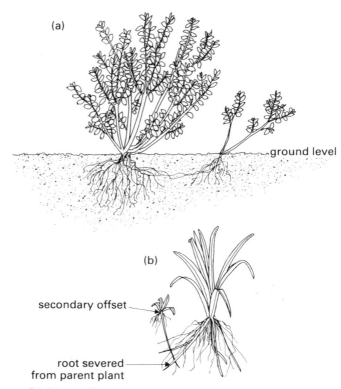

FIG. 20 Division (suckers). (a) A sucker growing from the roots of a forsythia. (b) A sucker/offset from a *Hemerocallis*.

Tuberous begonias can provide one example of division by this method, large vigorous tubers only being selected for the treatment. Tubers are started into growth in the greenhouse in February as normal and grown on until the top growths are 1.3 cm or so long. The tubers are then carefully lifted and cut vertically through the middle with a sharp knife. A light dusting of powdered lime is then applied to the wound to discourage harmful organisms and prevent bleeding and the separated sections replanted. It is best after this treatment to place the plants in a warm propagating frame to establish them quickly.

Dahlia tubers, although looking much different from the tubers of the begonia, are again divided in a similar manner (Fig. 21).

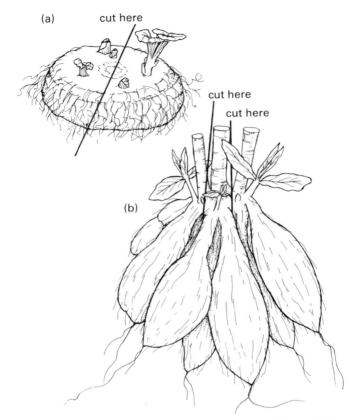

FIG. 21 Tubers. (a) Begonia tuber just starting into growth. (b) Dahlia tuber. Single stem tubers can also be cut by splitting through the centre of the stem.

Once more choose only the largest and strongest for the treatment. As with the begonia, the dahlia tuber is cut vertically through the crown of the plant making sure that each section cut away retains at least one growing bud or shoot and a section or finger of the original tuber.

SUITABLE PLANTS FOR THIS METHOD

Season	Plant	Type of division	Special requirements
LATE WINTER	*Cymbidiums* (Orchids) Garlic	 offsets	indoors outdoors
SPRING	*Antennaria* (Mountain Everlasting)		outdoors
	Arundinaria (Bamboo)		outdoors
	Asparagus Fern		indoors
	Aspidistra (Cast Iron Plant)		indoors
	Aster (Michaelmas Daisies)		outdoors
	Astilbe (False Goatsbeard)		outdoors
	Border *Campanulas*		outdoors
	Camptosorus (Walking Fern)	of natural layers	indoors
	Chives	offsets	outdoors
	Cornus (Dogwood)	suckers	outdoors
	Crocosmia (Montbretia)	offsets	outdoors
	Davallia (Hare's Foot Fern)	of rhizome	indoors
	Galanthus (Snowdrop)	offsets	outdoors
	Geranium (Cranesbill)		outdoors
	Geum (Avens/Herb Bennet)		outdoors
	Hemerocallis (Day Lily)		outdoors
	Hosta (Plantain Lily)		outdoors
	Nephrolepis (Ladder Ferns)		indoors
	Populus (Popular)	suckers	outdoors
	Rhubarb		outdoors
	Scabiosa (Scabious)		outdoors
	Sempervivum (Houseleek)	offsets	outdoors
LATE SPRING	*Calla* (Bog Arum)		outdoors
	Caltha (Marsh Marigold)		outdoors
	Glyceria		outdoors
	Hottonia (Water Violet)		outdoors
	Nuphar (Common Spatterdock)		outdoors
	Nymphaea (Water Lily)	of rhizome	outdoors
	Typha (Reedmace)		outdoors

SUMMER	*Aechmea* (Urn Plant)	offsets: suckers	indoors
	Aloe	offsets	indoors
	Androsace	offsets	outdoors
	Bryophyllum (Kalanchoe)	offsets: plantlets	indoors
	Chionodoxa (Glory of the Snow)	offsets	outdoors
	Colchicum (Autumn Crocus)	offsets	outdoors
	Iris	rhizomatous types	outdoors
	Narcissus (Daffodils)	offsets	outdoors
	Rebutia	offsets	indoors

AUTUMN	*Allium*	offsets	outdoors
	Aubretia (Rock Cress)		outdoors
	Bergenia (Bear's Ears)		outdoors
	Cerastium (Snow in Summer)		outdoors
	Gladiolus (Gladioli)	offsets	cold frame
	Helleborus (Christmas Rose)		outdoors
	Hyacinthus	offsets	outdoors
	Lilium	offsets: bulb scales	propagator
	Muscari (Grape Hyacinth)	offsets	outdoors
	Primula (H.P. Types. Primrose/Polyanthus etc.)		outdoors
	Raspberries	suckers	outdoors

4 LAYERING

Perhaps the best example of natural layering is the rooting of strawberry runners, or 'stolons'. All through the summer months the plants send out these runners from the axils of the leaves and they can cover quite a distance before the terminal bud produces a cluster of leaves and takes root. From the base of these new leaves arises another runner which moves on and so spreads or propagates the original plant rapidly and over a wide area (Fig. 22).

The gardener can take advantage of this natural method of propagation in the strawberry to increase his stock of plants. If a plant is required in an existing bed it is a simple matter to encourage a runner in the right direction and peg down the terminal bud where required. A few weeks later it will have sufficient roots to fend for itself and can be separated from its parent with a sharp knife, at the same time removing any runners forming from the new leaves.

Alternatively, a container of compost can be placed beneath a terminal bud and the bud pegged down to that. Eventually the new plant can be severed already potted up. As with all propagation material it is essential that the parent plant is healthy and free from disease.

To make an artificial layer similar to that of the strawberry is simplicity itself, nothing being being needed in most cases other than a wooden or wire peg, a sharp knife and a suitable subject.

Very many shrubs such as magnolias and rhododendrons are likely candidates for layering, along with many plants in the rock garden. The method is also particularly useful for those plants not propagated easily by other methods, as the prospective new plant remains connected to its parent until rooting is well established.

During July and August a young, non-flowering branch should be selected. In the case of shrubs this will naturally be one close to ground level.

At a point where the branch will easily reach the ground, the sap is restricted either by giving the stem a twist, bending it or preferably making an incision with a knife just below a bud. The cut should be made towards the main stem 2.5-7.5 cm in length and just deep enough to form a tongue. Do not cut too deeply or the stem will probably break later. Before pegging down remove

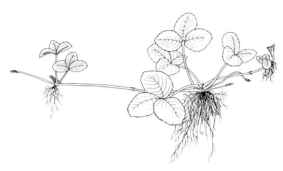

FIG. 22 Strawberry runners.

any leaves which may be beneath ground level when the layer is made.

In soil of good condition it might only be necessary to peg the branch down to the ground, keeping the tongue as open as possible without causing the branch to break, and then covering the cut with 2.5 cm or so of soil or compost. To assist in a satisfactory angle where the stem bends in the ground the leading edge of the branch can be tied in place to an appropriately placed cane.

If in doubt of the quality of the soil then a layer of some 5 cm of J. Innes Seed Compost or similar can be built up around the branch to be layered and the branch pegged into this, again covering with a further layer of compost over the top.

When sufficient roots have formed the new plant can be cut away from its parent. Root formation takes a variable amount of time depending upon the subject. Most shrubs take from six to twelve months, although rhododendrons and magnolias are better left for two years before separation.

Border carnations, which are also propagated by this method, should root and be ready for transplanting about eight weeks from being layered (Fig. 23). In all cases, should doubt exist as to the quantity of roots, leave the new plant attached to its parent for a further length of time. Shrubs, for instance, would benefit greatly if left to overwinter attached to their parent, separation and transplanting being left until spring.

AIR LAYERING

Now we come to a more specialised form of layering, originally

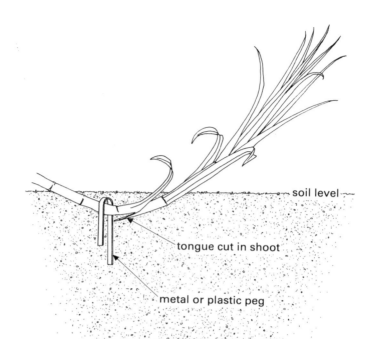

FIG. 23 Layering a border carnation.

introduced from China and still widely known as Chinese layering. It is useful for propagating a wide range of shrubby plants and trees and enables roots to be formed on stems well above the ground without using soil at all (Fig. 24).

For all purposes the method of operation is the same. The materials required are a few handfuls of damp sphagnum moss (soaked in water and squeezed out), some waterproof self-adhesive tape and a tube or sheet of thin polythene.

Select a good, non-flowering shoot and make an upwards slicing cut into the stem where you wish the roots to form. The cut should be from 2.5 to 7.5 cm in length depending on the size of the shoot, and never deeper than half-way through. Should there be any danger of the stem breaking, then a short length of cane tied to the stem above and below the cut will act as a splint to prevent

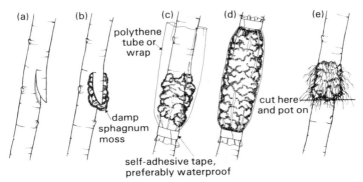

FIG. 24 Air layering. (a) Cut a tongue in the branch where roots are required to form. (b) Keep the tongue open with a little sphagnum moss. The cut in the stem can be treated with a rooting compound at this stage if desired. (c) Either surround the tongue with moss or slide a polythene bag over the branch and fasten in place at the base before packing tightly with moss. (d) Use a stick to push moss down into the polythene tube, and seal with tape at the top. (e) When roots appear on the side of the polythene, carefully remove the bag and any loose moss and separate the new plant by cutting off below the roots.

this happening.

Take a small amount of sphagnum moss and carefully insert it in the cut to keep it open, being careful not to break the stem. Dusting the cut prior to mossing with a hormone rooting compound may speed up rooting but is not essential.

With more moss build up a sausage shape around the cut, making the 'sausage' a good handful of moss. With outdoor work, especially, wrapping the moss around with cotton can save some frustration.

The final stage is to wrap the moss in polythene to exclude excess water when layering in the open air, and to retain moisture in all events. The thinner the polythene the better. With the polythene wrapped around the moss it is held in place and sealed with tape at the top and bottom in such a way as to prevent rainwater seeping in. Waterproof tape is, of course, essential.

If a polythene tube is used a slightly different procedure can be adopted, the tube being first slipped over the top of the stem and secured with tape below the cut. The tube is then packed tightly with moss, pressing it into position with a short stick or cane, and the top sealed as before.

After a period of time, dependent on the subject, the time of year etc, roots will be seen forming against the side of the polythene. When these appear well established the polythene can be removed and the stem severed below the sausage of moss.

Carefully remove all the loose moss but avoid damaging the roots and simply plant or pot the newly-rooted plant into its correct environment.

Air layering is usually undertaken in spring or early summer when best results can be expected. As a general guide such plants as *Ficus* species (rubber plants) will form roots in some six to eight weeks under glass, whereas outdoor shrubs may take in excess of ten.

SERPENTINE LAYERING

This is a method particularly suited to long shoots such as those found on clematis and lonicera (honeysuckle).

Slanting cuts some 5 cm long are made with a sharp knife behind nodes at convenient distances along the stem chosen for propagation. These cuts are then pegged down into pots of sandy compost or directly into the ground, the intervening lengths of stem being left in the open. The pegged-down sections are covered with a further layer of compost and left to root (Fig. 25). Clematis is normally propagated by this method in June, roots forming by autumn when the new plants are separated from their parents and either potted singly for protection through the winter or planted straight into their new growing positions.

TIP LAYERING

As with the layering of strawberries, this is a method exploiting a natural method of propagation, and it is very simple to undertake.

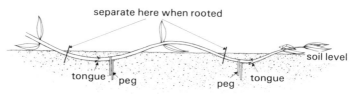

FIG. 25 Serpentine layering.

It is used widely for propagating cultivated blackberries and loganberries, the tips of strong growing shoots being pegged down or dug into the ground where they touch. Roots will then form from the tip and by autumn the shoot can be severed and transplanted. Rambling roses, currants and gooseberries can also be propagated using this method.

SUITABLE PLANTS FOR THIS METHOD

Season	Plant	Type of layering	Special requirements
SPRING	*Ginkgo* (Maidenhair Tree)	air	outdoors
	Hamamelis (Witch Hazel)	air	outdoors
	Hydrangea petiolaris (Climbing Hydrangea)	Serpentine	outdoors
	Jasminum (Jasmine)	Serpentine	outdoors
EARLY SUMMER	*Clematis*	Serpentine	outdoors
	Ficus (Rubber Plant)	air	indoors
	Magnolia	air	outdoors
	Parthenocissus (Virginia Creeper)		outdoors
	Wisteria	Serpentine	
SUMMER	*Alnus*		outdoors
	Blackberry (Rubus)	tip	outdoors
	Border carnation		outdoors
	Chimonanthus (Wintersweet)		outdoors
	Davidia (Handkerchief Tree)	ordinary or air	outdoors
	Rhododendron		outdoors
	Strawberries		outdoors

5 GRAFTING

As a means of propagation, grafting is rather slow and time-consuming, but if it was not for this method many of the plants we grow, whether in private gardens or commercial holdings, would not give the cropping results they do, indeed some would undoubtedly fail completely.

The principle of grafting is that of joining two parts of a living plant together so that they form a permanent union. One part of the plant (the rootstock) provides the eventual grafted plant with a strong root system and main stem, and the other part (the scion) forms the fruiting/flowering section of the plant. The scion is usually a section of stem taken from the previous year's growth of the parent plant.

By doing this, the fruiting qualities, for grafting is mainly used for the propagation of fruit trees, of choice 'cultivated' varieties can be combined with the rooting strength and vigour of the wild species or their selected strains.

As with the ever-increasing incidence of organ transplants in our human species, the two key words for successful grafting are compatibility and cleanliness. It therefore goes without saying that the rootstock and scion should belong to the same genus or plant group, although in a few cases plants within the same family can be successfully grafted together. It is also essential that both rootstock and parent plant should be disease-free and of vigorous growth.

There are many methods of making a graft and a great many more variations on those initial methods, but whatever graft is used the basic physiological principle remains the same: the cambium layer (*see* page 78) of the scion must come into direct contact with the cambium of the rootstock.

When the unions are made they are normally bound in position with raffia, or one of the more modern rubber, plastic or fibre bands specially manufactured for the purpose. In most cases, after binding, the union is sealed with grafting wax or a bitumen emulsion which serves the same purpose. Under glass, petroleum jelly serves reasonably well.

The grafting of fruit trees and ornamental shrubs is usually undertaken at the end of the dormant season in early spring. The scions are collected and prepared earlier than this, in December or

January, tying them in bundles and storing them in cool soil against a north wall or fence. Commercially, special bins are made for this. In this way the growth of the scion is retarded until the sap is moving freely in the rootstock.

Scions are chosen from good stems of the previous year's growth and are cut from the parent tree at a length sufficient to contain four or five good buds. Before storage the soft tips are removed from each stem.

When the time is right, the scions are taken from their store and cut down to lengths of three buds. The preparation after this depends on which method of grafting is being used and these are outlined later under their various titles.

Fruit trees and ornamental shrubs are not the only subjects suitable for grafting methods and many woody perennials can be grafted along with such somewhat unlikely candidates as cacti and certain houseplants.

The methods of making a graft are so numerous and for so many purposes that I have dealt only with those most commonly in use and in my opinion closest to the propagating function. Few tools are needed for those grafts outlined here and most will be in the gardener's tool shed anyway. A strong sharp knife is essential for much of the work, with razorblades being suitable for the softwood candidates. Pruning saws and secateurs are useful in some cases.

SPLICE GRAFT

An ideal graft for the beginner, this is probably the simplest method of all. It is much used for grafting fruit stocks where scion and rootstock have a similar diameter. Its one limitation, however, is the need to hold them in position while the tie is made, there being no mechanical connection between the two parts.

Both stock and scion are prepared with a long slanting cut and the two cut surfaces brought together. Care should be taken to ensure that the cambial regions are in contact before finally tying and sealing.

Broom, roses and clematis can also be grafted using this method.

WHIP AND TONGUE GRAFT

This is a very similar method to the splice graft but with the

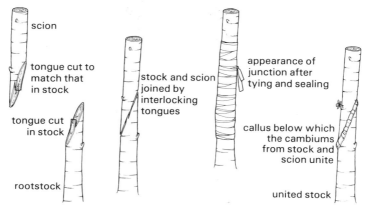

FIG. 26 Whip and tongue grafting.

refinement of a tongue of wood cut in both partners to hold them together while a tie is made.

The method is best suited to stock and scions of the same diameter, particularly those less than 2.5 cm. The scion is first cut with a slanting cut from its base extending some six times its diameter. A second cut is then made towards the top of the first in such a way as to form a downward pointing tongue. The stock is then prepared similarly with a corresponding slit made so that when together the two tongues interlock. The graft is finished with tying and sealing (Fig. 26).

This method is much used in fruit tree production in spring using retarded scions as described earlier. As with all grafts, any ties made are released by cutting once a satisfactory union is seen to have been made.

WEDGE OR CLEFT GRAFT

The wedge graft has widespread uses and in its simplest form is easily made. The rootstock is prepared by cutting it off at an appropriate place and making a downward cut into which the scion is placed. The scion in its turn is cut to a thin wedge at its base and slid into the downward cut in the rootstock making sure there is cambial contact at one side at least. The joint is lastly tied with raffia, tying the scion down with a loop over the first leaf if it shows signs of being forced out by the tying process (Fig. 27).

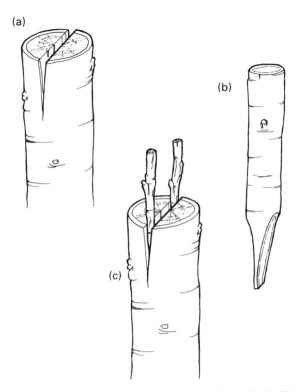

FIG. 27 Wedge or cleft graft. (a) A cleft cut across the stock. (b) The scion cut to a wedge shape. (c) Two scions placed opposite in the cleft.

The method is also used in its most drastic form in topworking trees when the appropriate wedge slits are made in the stock with a grafting tool or axe. The variation known as oblique wedge grafting is to be preferred in this case, the slits in the stock being made in such a way that they do not extend fully through the branch as would be the case with the true cleft graft.

Where herbaceous or in-leaf scions are used the grafted plants should be confined to a warm close atmosphere until a union has been made.

SADDLE GRAFT

This method is often used for grafting plants which are

difficult to propagate in any other way, such as certain forms of rhododendron which are grafted onto rootstocks of *Rhododendron ponticum* (Fig. 28).

For a successful union it is important that the stem of the scion and rootstock should be much the same size, the rootstock for the purpose usually being a two or three year old seedling grown on in a pot for its last year in the case of rhododendrons.

The base of the scion is prepared by making two cuts slanting inwards to produce an inverted V, the knife being turned sharply at the end of each cut to form a kind of saddle. The stock is then worked in a similar way to take the saddle, the two parts being matched as closely as possible. Again, in the case of rhododendrons, the graft is made as close to the roots as possible so as to be underground when eventually planted out. In this way

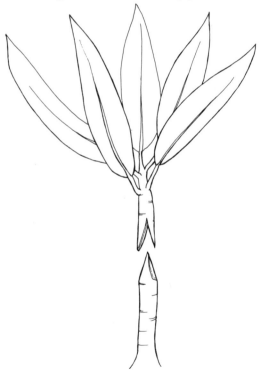

FIG. 28 Saddle grafting.

the scion can form some roots of its own, desirable in this case.

The graft is generally done under glass and as with other subjects done in this environment the grafted plants are best placed in a warm close atmosphere until a union has been made.

APPROACH GRAFTING

This method is in effect only an artificially induced 'natural' graft, the same as can be found on ivies growing wild where two stems crossing one another have grown together during the years. The method does differ from other forms of grafting in that the stock and scion are joined and established before the scion is severed from its own roots.

In its basic form the stock and scion are generally grown in separate pots and the two brought close together when the graft is to be made. With the two plants as close as possible, an appropriate place on the stem of each plant is chosen and a shallow sliver of bark and wood is removed from each, thereby exposing the cambium layer beneath (Fig. 29).

The sliver should, as far as possible, be much the same size from each stem so that there will be a good join between the two. As both plants are on their own roots, the graft can be done at almost any time, but is best when the sap is running freely in spring or summer.

With the two cuts made, the stock and scion are brought together and tied in place (Fig. 29), making sure that the cuts

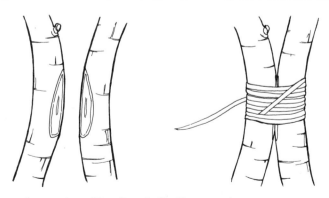

FIG. 29 Approach grafting. Detail of knife cuts and union partly tied in position.

match as closely as possible. When the union has been formed satisfactorily, the top of the rootstock is cut away above the graft and the root of the scion below, this being done gradually over a number of weeks to lessen the shock to the plant.

Vines can be grafted by this method and it is also useful for subjects difficult to graft as both sections are able to sustain themselves on their own roots until a union is made. This type of graft is usually done under glass.

FRAMEWORK GRAFTING

Framework grafting is a general term which covers a number of grafting methods used to rework or convert the head of established trees from one variety to another. It differs from 'topworking' in that the main framework of the tree is retained, only the lateral shoots and spurs being replaced. It is the better of the two methods, frameworked trees coming into fruit production in less time than those which have been topworked.

To prepare a tree for frameworking, the side branches should be thinned out if necessary to shape the tree nicely and then all the smaller laterals and spurs are pruned away completely from those remaining, except where the tree is to be worked by stub grafting. These branches are then replaced by scions of the required variety, making the grafts every 20-25 cm apart (Fig. 30).

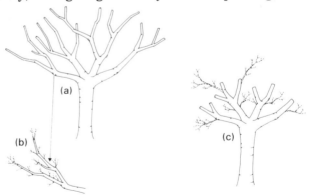

FIG. 30 (a) An apple tree prepared for frameworking. (b) Detail of a worked branch with the scions shown in dotted lines. (c) The same tree prepared for topworking. The sap-drawing branches are indicated by dotted lines.

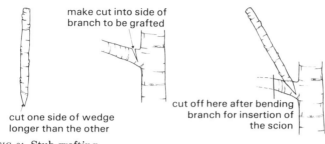

FIG. 31 Stub grafting.

The scions used are longer than normal, bearing a minimum of six buds or as many as seven or eight. As the object of frameworking is to make the tree appear as natural as possible great care should be taken in placing the scions, remembering that scions grafted vertically on the inner or top surface of a branch will grow stronger than those in a horizontal position below.

Various methods of grafting are used in frameworking, stub grafting being one (Fig. 31). With this, the tree is prepared as before except that lateral growths of between 6 and 25 mm are left intact wherever a new growth is required.

The scions are prepared with a wedge-shaped base, one side being slightly longer than the other. A cut is then made in the branch to be grafted some 1 cm away from the main stem, making the cut towards the base of the branch but never deeper than its centre.

The branch is bent slightly to open the cut and the wedge of the scion placed in it, longer side of the wedge down. Once released the branch will hold the scion tightly, making it unnecessary to tie it in place though a few turns of tape around the graft would do no harm. With the graft made satisfactorily the original branch is cut away immediately beyond the graft and the job completed with a sealing compound. When the tree is finished the terminal branch on each main branch should be removed immediately above the topmost scion.

If stub grafting alone is insufficient to give the tree a balanced framework the gaps can be filled using a method called side grafting. This is accomplished by preparing the scion with a wedge-shaped base, one angle of the wedge being more acute than the other. This wedge is then inserted into a cut made in the side of the branch, the cut going a quarter of the way through. Both wedge and cut are made in such a way that, when finished, the scion will protrude from the branch at as natural an angle as possible.

Once again the cut is opened for insertion by bending the branch slightly so that when released it will grip the scion tightly.

Other methods used in frameworking include bark grafting (Fig. 32) and oblique side grafting, and in all cases methods can be mixed if doing so will yield a better overall result.

It should be remembered that more than one variety can be worked on one tree, which is useful when cross-pollinators are required. It is useful, too, for those species having male and

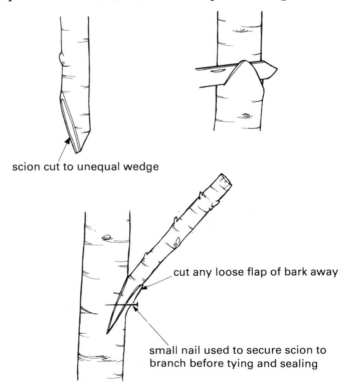

FIG. 32 **Bark grafting.**

female forms, such as *Ilex aquifolium*, the common holly. This same principle is followed when two or more varieties of apple etc., are budded or grafted onto a single rootstock to create a 'family' tree.

CROWN OR RIND GRAFTING

This method is used mainly for the topworking of established trees when it is desired to change one variety to another. It is best carried out in late April in most parts of the country, the scions being prepared earlier and stored and the tree to be worked being prepared by cutting away most of the main branch system some 60-90 cm above the crutch (the part of the tree where the branches separate from the trunk). A few smaller branches are left just below the main level of grafting to act as sap-drawers. These feed the roots and sustain the tree until the new branch system is established (Fig. 33).

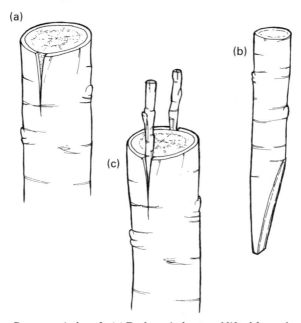

FIG. 33 Crown or rind graft. (a) Bark or rind cut and lifted from the wood. (b) A scion cut to fit is pushed between bark and wood. (c) Two scions placed opposite.

The larger the tree the more sap-drawers are retained, these being removed or worked in their turn when the new wood is growing well.

Where possible, branches should be removed above a fork, making for more grafting, but lessening the chance of infection. For this same reason, the tree is best prepared and grafted in one operation.

Before commencing the graft, the cut on the branches to receive the scions is pared smooth with a knife, and the scions are then inserted under the bark or rind at the top of the branch. To do this, the scion is prepared with a flat slanting cut at its base, extending some six times longer than the diameter of the scion. A corresponding cut is then made down the side of the receiving branch from the top and the scion is pushed down between the rind and the wood, the cut surface of the scion facing inwards (Fig. 34).

A varying number of scions are inserted into a branch depending on its diameter, two to four being the general rule. Once made, the graft is bound and the whole wound encased in grafting wax or similar.

Many other methods are used for topworking other than the crown graft, cleft and veneer grafting being among them.

Topworked and frameworked trees should not be neglected

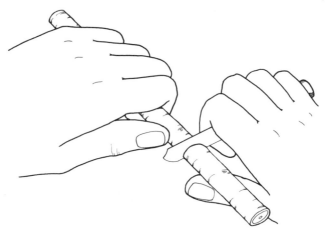

FIG. 34 Position of hands when cutting a scion to fit the cut in the stock.

after grafting and must be treated in much the same way as fruiting trees to keep them free from pests and diseases. Care in the early stages will pay dividends. In situations exposed to strong sun it may also be advisable to whitewash the trunks of topworked trees to prevent damage by strong sunlight, a condition known as sunscald.

As might be expected with such drastic attention the trees will throw out many suckers and most should be removed, especially when they appear around the scions. A few can be retained and will act as sap-drawers for the first season or two. As with all fruit growing, attention to good pruning in the following years is necessary for good results.

DOUBLE-WORKING

Although not likely to be used by many amateur gardeners, this is still an interesting method of grafting to consider as its application is quite widespread.

Double-working is used for various reasons. The main ones are to overcome incompatibility between stocks which in theory should be compatible, and to build trees with strong straight stems, a process known aptly as 'stem-building'.

The first situation is commonly found with pears which are grafted onto quince rootstocks. Some varieties take well to the graft, but others, though perhaps developing an effective union for growth purposes, are weak in structural strength and subsequently break off as the tree becomes established. To overcome this an intermediate scion known to be compatible with both partners is grafted between the two, hence the term 'double-working'.

Even though this intermediate stem may only be a few centimetres long it can have a profound effect on the finished tree, and this effect has been successfully used to produce disease resistance and frost hardiness in top scions not known for these qualities.

Double-working can be used to produce straight stems when the variety you wish to grow is inclined to produce a bush rather than the desired tree. Using a mutually compatible scion of a good stem-producing strain, the intermediary is grafted to the rootstock first and the top scion added at that part of the stem where the head of the tree is desired to form.

Double-working is usually carried out in one operation, the scions and stock being prepared in the normal way. The intermediate scion is cut about 13 cm long and is grafted to the usual three-bud top scion. Whip and tongue grafting is perhaps best for this. After tying and sealing, the intermediate and top scion can be grafted onto the prepared rootstock using the same method.

Shield budding (*see* Chapter 6) can also be double-worked when budding pears, a small budless shield of an intermediate being

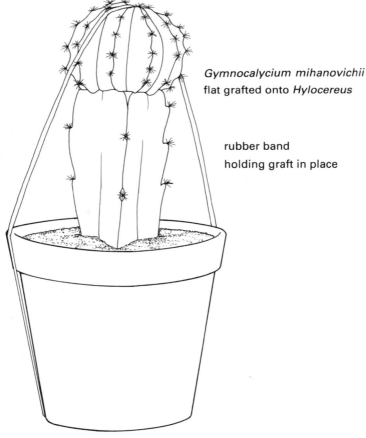

Gymnocalycium mihanovichii flat grafted onto *Hylocereus*

rubber band holding graft in place

FIG. 35 Flat grafting a cactus.

inserted between the rootstock and the true bud shield as the operation is done.

FLAT GRAFTING

Grafting is used on a number of cacti for a variety of reasons, the main one being to speed on the growth of slow-growing species or to create a 'standard' with a naturally pendulous species grafted onto an upright stock. The same principles of cambial contact and cleanliness apply to grafting cacti as with other plants and it is necessary to work fairly quickly to avoid the cut surfaces drying out before contact is made.

A flat graft is most often used for the former reason, and plants can be grafted from seedlings one year old onwards, ideally using rootstocks and scions of the same diameter. They are simply cut across straight at the required spot with a razorblade and fitted together. Hold in place either with a couple of rubber bands or with a cactus spine pushed through the side of the scion, the former being preferable, in my opinion, as it causes no damage if done correctly. Use a small cloth pad on top of the scion if necessary.

VENEER GRAFTING

Although many conifers can be propagated from seed and/or cuttings, cultivars proving difficult or impossible by these methods have to be grafted. Among these are varieties of *Cedrus* (Cedars), *Cupressus* (Cypress), *Picea* (Spruce), *Pinus* (Pines) and *Pseudotsuga* (Douglas Fir), many of which respond well to the veneer graft.

This is a graft in which the top remains on the potted two-year-old rootstock until a union is made. In the case of conifers, the scion is a piece of the previous year's growth, usually between 5 and 10 cm long, depending on the speed of growth of the donor. The slower the growth the shorter the scion.

The rootstock is prepared with a downward vertical cut just below the bark about 2.5 cm long and as near the compost surface as possible so as to produce a loose flap of bark with the wood just visible below. Cut the scion then with an equal-sided wedge cut at the base exactly matching the cut in the rootstock, then push the wedge down behind the flap making sure that there is cambial contact on at least one side.

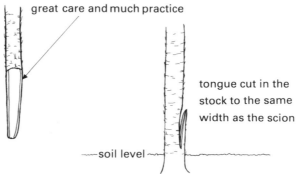

FIG. 36 Veneer grafting.

With the scion in place, bind the wound with raffia and place the pot with rootstock and attached scion into a propagator with bottom heat until the union is formed. This takes some six weeks or so, and after that the raffia can be removed and the rootstock beheaded before hardening the new plant off to outdoor conditions.

ROOTSTOCKS

A description of all the rootstocks and their various methods of propagation and their specific uses would take a chapter of its own, so this section contains merely an introduction and one method of propagation peculiar to rootstock production.

Rootstocks can be raised in many ways: by seed, cuttings, layering and stooling.

Most methods are outlined elsewhere, but stooling is dealt with here. With this, a parent rootstock is planted and grown on until it is well established, then during the winter it is cut off near the ground. Treated in this way it responds by sending out a number of shoots from its base. When these are 12-15 cm high they are earthed up, continuing throughout the year as the shoots grow until the shoots are buried in 15-20 cm of soil. This keeps all light from their bases and they form roots during the summer.

In late autumn the soil is removed and the rooted shoots are broken away from the original rootstock now called a stool. Plants treated this way will yield a crop of shoots annually.

Much work on rootstocks has been done at the East Malling Research Station in Kent and the strains developed there and jointly with the John Innes Institute at Merton (before it moved in 1949) are used generally for the production of fruit trees. Taking apple stocks as an example, these are identified with numbers corresponding to their growth characteristics such as M (Malling) 9 and MM (Malling/Merton) 106, both rootstocks producing dwarf trees. Other rootstocks produce medium-sized and vigorous trees.

Prior to working, most hardy fruit stocks are planted out in the spring of one year and are grafted in the March or April of the next. It does no harm at all to leave them an extra year if they do not appear to be strong enough. Plant out leaving 38 cm between the plants in rows 90-120 cm apart if your planting extends to such numbers.

The whip and tongue graft is most used in this situation, the rootstocks being cut down to within 7.5-10 cm of the ground at the time the graft is made.

SUITABLE PLANTS FOR THIS METHOD

Season	Plant	Type of graft	Special requirements
LATE WINTER	*Apples*	whip and tongue	outdoors
	Clematis 'Jackmanii'	cleft	under glass
SPRING	Conifers (varietal forms)	veneer	under glass
	Hardy trees (varietal forms)	whip and tongue	outdoors
	Pears	whip and tongue	outdoors
	Rhododendron	saddle	under glass
SUMMER	*Gymnocalycium*	flat	propagator
	Paeonia	wedge	cold frame
	Schlumbergia	cleft	propagator

6 BUDDING

The method of propagation known as budding is really a form of shield grafting. It has, however, developed into a branch of propagation in its own right and is extensively used for the propagation of roses, although many fruit trees and ornamentals are also increased by the method. The technique used is much the same in all cases, the only real difference being the height at which the bud is placed on the stem of the rootstock, so I have chosen to describe in detail only one example, that of budding roses.

Although some roses, particularly those nearest the original species (i.e. *Rosa moschata*, the Musk Rose), can be raised from cuttings, many modern varieties or cultivars are not strong enough to perform on their own roots to any satisfaction. To overcome this problem such varieties are budded onto various strong growing rootstocks which, generally speaking, are wild species of rose or selected strains of wild species. The object of the graft is to transfer a well-developed bud from the cultivated variety to the rootstock.

The main method used in the budding operation is quite simple to accomplish, although as with many things only practice makes perfect.

The first essential is to undertake the graft when the plants are in full growth. The sap must be travelling freely within the stem to allow the bark to rise from the wood of the rootstock without damaging the vital cambium layer, this being the narrow layer of growing tissue between the bark and wood of most plants. It is this cambium layer which is responsible for healing wounds, forming callus on cuttings prior to rooting, and for the uniting of budded or grafted stocks. In normal seasons plants are in full growth between mid-June and the end of August. Obviously those gardeners in the more southern areas will have a longer choice of season than those in the north.

The only tool needed for successful budding is an extremely sharp knife. Special budding knives can be bought and are best suited to the job. These knives have a single blade no longer than 5 cm in length and a 'bone' handle, the handle being tapered to a flat point to assist in the raising of the stock's bark.

As I have said, a knife designed for the purpose is naturally the best tool for the job, but as long as the operation is carried out with care any knife of a similar design can be used. A suitably shaped piece of wood or plastic can be used to lift the bark, or handymen may even be able to modify or manufacture a budding knife of their own. A typical design is shown in Fig. 37.

The shoot selected to supply the bud should be a good one which is carrying a flower near to shedding its petals. The shoot should be cut some 15-20 cm in length so that it carries several buds or 'eyes'. These eyes are identified as small knobs in the junction where the leaf stalk joins the stem. The eyes should be plump and firm but not yet started into growth, and the best eye will probably come from the centre of the cut stem.

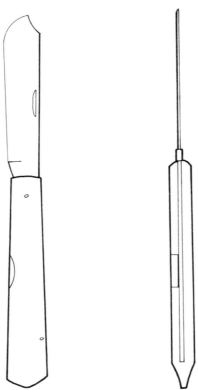

FIG. 37 A typical budding knife.

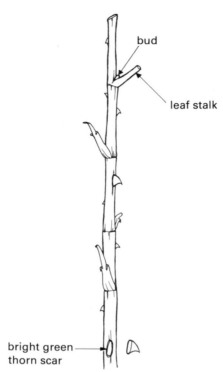

FIG. 38 A prepared bud stick.

The thorns on the shoot are also good indicators as to the hardness of the stem and can be used in determining the readiness of the buds for budding. The idea is to try to break off the larger thorns by pressing them sideways with the fingers (Fig. 38). If the thorn tears off the stem with shreds of bark attached it is not ready and should be left another week or so. If, on the other hand, they snap off cleanly, the stem is ready to use. The scar left by the removal of the thorn should be of tender green tissue. An overripe scar will appear hard-looking and dry. A stem of this type is also unsuitable.

Once a stem is selected it can be removed from the parent plant at the same time cutting the leaves off, leaving 2.5 cm of leaf stalk attached to the stem to protect the bud and allow for easier handling. At all stages drying out of the stem is to be avoided and prepared stems should be placed in a glass or jamjar containing a little water to prevent this.

The next stage is to prepare the rootstock. This is done by removing the soil from around its stem as close to the roots as is practicable and wiping the resultant exposed area with a rag to clean it.

A T-shaped cut is then made just below the original ground level (Fig. 39). This cut should be no deeper than the bark, with the downstroke of the T-cut being about 2.5 cm in length. Once the cut is made the bark is raised carefully from the stem to form two flaps. If the rootstock is in prime condition the bark will lift cleanly. On no account should it be forced as this would damage the cambium layer and possibly cause failure of any graft made.

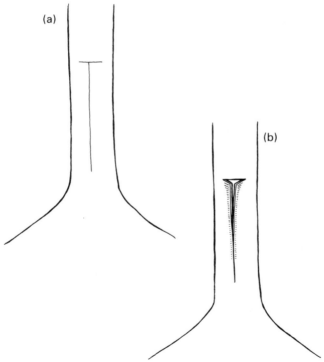

FIG. 39 (a) Position of the T-cut on the rootstock. (b) Bark raised to form two flaps.

With the T-cut made, press the flaps back against the stem to avoid any drying while the bud itself is prepared. This is done by cutting the bud or eye from the prepared stem complete with a shield of bark. Start the cut about 2.5 cm below the chosen bud and gradually cut deeper until the bud is reached, whereupon the cut is made shallow again to allow the knife to emerge around 2.5 cm above the bud and, naturally, on the same side of the stem. What is then produced is a shield of bark containing a bud, a leaf stalk, and at the back a thin sliver of wood (Fig. 40).

The next step is perhaps the most difficult as the sliver of wood has to be removed from the shield without causing any damage to the root of the bud behind it. The beginner would perhaps be best

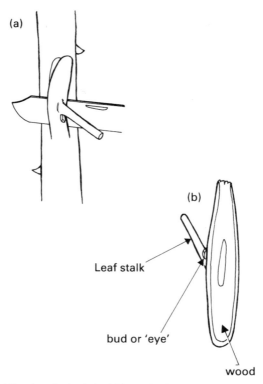

FIG. 40 (a) Cutting the bud shield from the bud stick. (b) Detail of bud shield.

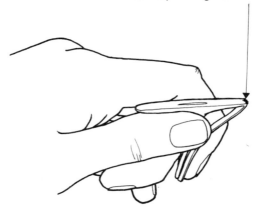

FIG. 41 Removal of wood from the bud shield.

advised to practise on a few rejected buds which can be thrown away with no loss before the real buds are attempted.

There are a number of methods of removing the wood from the shield, but the one I prefer is used as part of the entire operation. It is effected by making the cut as before, except that the knife is withdrawn before the blade reaches the end of cut and the job is completed by tearing. This results in the shield terminating in a long strip of bark. The wood is then removed in two stages. Firstly, it is loosened from the long tail of the shield by waggling the latter to and fro for a moment. Next, the shield is held by the leaf-stalk between the thumb and index finger of the left hand, the cut surface uppermost and the tail of the shield lying along and in line with the thumb (Fig. 41), the second finger of the hand holding it in place. This holding will cause the wood to rise slightly from the tail so that it can be grasped by the fingers of the other hand. By flickering the wood upwards and backwards and increasing the bending motion as the bud is reached, the wood is removed.

With the wood gone, the bud root should appear as a definite plumpness behind the bud on the shield. If, however, the area appears hollow it will mean that the root of the bud has been torn out with the wood making the shield useless. A few experiments will enable the beginner to judge what is correct. Take heart from the fact that the written instructions are far more complicated than the actual handling of the plant material!

After giving these instructions it has to be said that not all growers advocate the removal of the wood from the shield, saying that its retention only rarely results in a failure of the graft. This may very well be so, but if you should decide to leave the wood on the shield it is essential to make your cut below the bud as shallow as possible so reducing the thickness of the wood behind the eye.

With the bud shield and rootstock now prepared, the base of the shield is trimmed with a cross-cut 13 mm below the bud. Once more lifting the bark at the T-cut in the rootstock, the shield is carefully pushed down the cut as far as it will go. For this purpose the leaf-stalk makes a convenient handle and also ensures that the eye is facing outwards in its correct position. With another cross-cut the top of the shield is trimmed to fit neatly in place at the top of the cut. The bark is then pressed into place around it. Finally, the wound is bound firmly yet not tightly with moist raffia, leaving only the bud and leaf-stalk exposed (Fig. 42).

After three weeks or so it will be seen whether or not the bud has taken. Both the bud and the shield should look bright green. If they look brown or black they have failed and another attempt will have to be made on the opposite side of the rootstock and preferably still lower down the stem.

If the bud has taken the tie should be removed and the plant left to grow on as it is until the following February. At this time the stock plant should be 'headed'. This means that all the top growth of the original rootstock is removed from about 13-25 mm above the budded eye.

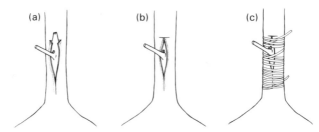

FIG. 42 (a) Bud and shield inserted in the rootstock. (b) Bud and shield in finished position. (c) Bud and shield fixed in position with moist raffia.

The new growth will sprout out later in the spring and it may be necessary to tie it to a small cane on the opposite side of the stock to encourage upright growth. New plants will flower some time later than established plants and after flowering should be cut back fairly low to encourage new growths from the budding point. The new plant can be transplanted to its permanent position in the autumn.

CHIP BUDDING

A brief mention may be made here of another method of budding which is said to be easier and just as successful as the traditional method. The operation involves the removal of a chip of stem from the rootstock and the replacement of this by a similar piece from the mother plant containing a bud, as with ordinary budding.

This is effected by making two cuts across the rootstock 25-40 mm apart and removing a chip of stem from between the cuts with a knife. This is replaced with a chip from the mother plant, the bud being near the centre of the replacement chip. When in place it should be tied securely with raffia as before.

ROOTSTOCKS

There are many different rootstocks which can be used for budding roses, the type chosen depending on the soil the plants will be grown in, the rose to be budded and the rootstock most readily available to the prospective budder.

These various rootstocks can be raised in two ways; either by cuttings or from seed. They can also be bought as maiden plants (one year old), from certain nurseries who specialise in rose-growing. Generally speaking, seedlings are to be preferred as they provide the better root system. For both methods of propagation, *see* Chapters 1 and 2.

The most readily available rootstock for the amateur is probably the wild dog rose, *Rosa canina*. Stock from this species, either from cuttings or seed, is best suited to growing on heavy soils. Commercially, many strains of *R. canina* have been selected, but if the species grows wild in your area it is a good indication that stock from such plants will do well in your garden also.

For light sandy soils, except those over chalk, it would be best to use *Rosa rugosa* or *Rosa multiflora japonica* (syn. *R. polyantha multiflora* or *R. polyantha simplex*), which also does well on shallow sandy soils. For light soils of a calcareous nature *Rosa laxa* is to be recommended. This species has a thin bark and is easy to bud; its main fault is its tendency to throw up suckers. (These can, however, be used to provide more rootstocks.)

Bush roses are budded onto two-year-old plants, mainly of *R. canina* or its numerous strains. Standard roses need to have a rootstock with a stem some 120-150 cm in height. It is possible to find such stems growing as suckers from hedgerow plants in the wild, but commercially standards are budded onto stems of *R. rugosa* or *R. polyantha simplex*.

With standards, the method of budding is the same as for bush roses, except that it is carried out at the top of the stem rather than near the roots. Usually, three buds are inserted on either side of the main stem at some convenient height, in the case of *R.*

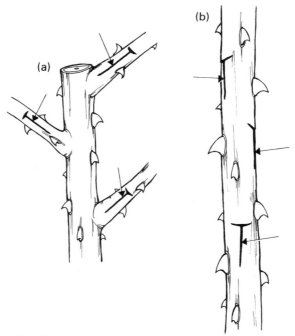

FIG. 43 Budding standard roses, showing the position of the T-cuts. (a) *Rosa canina;* (b) *Rosa rugosa.*

rugosa, or on the side branches growing from the three buds left to grow on the original cutting in the case of *R. canina* (Fig. 43).

When planting cuttings for rootstocks of either bush rose or standard production it is essential to remove all buds except the top three prior to insertion of the cutting. This will hopefully avoid a mass of suckers and unwanted sidegrowths on the eventual budded plant.

Rootstocks of all types intended for budding should be planted out in their budding positions in the autumn preceding the budding period. Needless to say, the area chosen for planting out should be in good light and the soil cultivated to a satisfactory growing condition.

SUITABLE PLANTS FOR THIS METHOD

Season	Plant	Special requirements
SUMMER	Apples	outdoors
	Apricots	outdoors
	Cherries	outdoors
	Ilex (Holly)	outdoors
	Peaches (of named varieties)	outdoors
	Pears	outdoors
	Roses	outdoors

7 GREENHOUSES, FRAMES AND PROPAGATORS

Without doubt, the owner of a heated greenhouse with a propagator and mist unit is by far the best placed to propagate the maximum number of garden subjects. Provided that he has the space, there is little he could not consider trying with a reasonable chance of success. Unfortunately, not everyone has these facilities, but they are by no means essential to an interesting propagating programme.

To begin, I will take as an example the lot of the average gardener whose interests and financial abilities evolve over the years. So we start with a house and a garden and little else except perhaps enthusiasm.

At once we can see that the range of plants which can be propagated without any protection at all is quite large. The seeds of most vegetables, hardy perennials, biennials and annuals are sown directly into the open ground during the warmer parts of the year. To these may be added various hardwood and softwood cuttings which will certainly root in a sheltered part of the garden, and among other methods of propagation layering, division, budding and grafting all have their subjects suitable for working in the open air, indeed the majority of such methods are carried out there.

Yet the average gardener still dreams of better things, so he buys a few cloches. These are best described as portable low frames and nowadays come in a variety of shapes and sizes from the bell jar to the polythene tunnel cloche. Many seeds can be sown much earlier if given cloche protection and either grown there to maturity or uncovered later as the weather warms. Cloches can also be used to protect hardwood and semi-hardwood cuttings as well as plants raised in the greenhouse and in need of hardening off to an outdoor climate (Fig. 44).

One step up from the cloche, though in no way superseding it as all additions to plant protection are complementary, is the cold frame. These are to be had in various proprietary designs, but the handyman can easily construct one of his own. The walls can be made of wood, brick or concrete, the top being covered with a frame light or lights of glass, polythene or clear plastic. The size is merely a matter of convenience to the individual, but a frame

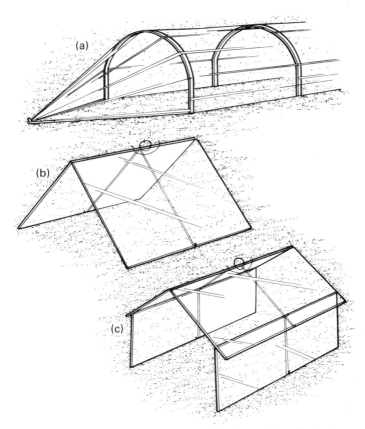

FIG. 44 (a) Polythene tunnel cloche. (b) Tent cloche. (c) Barn cloche.

1.5×1 m is a useful size. The walls are best made about 45 cm high at the back and 30 cm at the front.

As with greenhouses, frames should be sited in good light away from the direct shade of trees or buildings, though some measure of protection from cold winds is to be desired. Greenhouses are generally recommended to be erected to run from north to south, but it may be noted that a greenhouse or frame running east/west will receive more light in the winter period and so is best suited to propagation purposes.

The cold frame is a very useful addition to the garden either for raising plants in its own right or as a transit post for hardening off plants raised in a greenhouse.

The usefulness of a frame can be greatly increased by the addition of underground mains voltage heating cables which are available in various wattage ratings. Naturally an electricity supply to the frame is essential and should be fitted using outdoor grade cable and fittings, preferably by a professional electrician, but in all cases to a high standard of safety and workmanship.

The heating cable should be buried under some 5 cm of sand, being careful not to let it cross over itself or touch. As a rough estimate a frame will require a cable size calculated at 6 watts per 100 cm^2 of floor area plus a further circle of cable placed around the side of the frame just above the sand surface calculated at 15 watts per 100m^2 of glass or polythene above. For best results the cable is generally thermostatically controlled.

From the heated frame it is a simple but expensive step for a gardener to buy a heated greenhouse. This can of course be adapted for propagation only, when a greenhouse with half walls is to be preferred giving a better degree of insulation. Most gardeners, however, will want a greenhouse to serve a variety of purposes, and a propagating frame built within the greenhouse can serve them well.

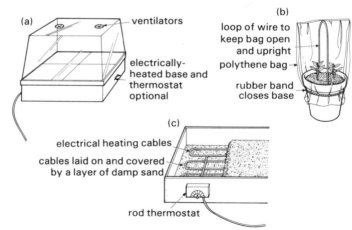

FIG. 45 Propagators. (a) A commercially-produced electric propagator. (b) A polythene bag propagator. (c) A home-made propagator base.

Heated cables buried in sand are best for the indoor propagator too and an area similar to the frame outside can be constructed, the materials needed naturally being much lighter. Careful construction will allow it to be removed or adapted when not in use for its purpose (Fig. 45).

An alternative is the professionally constructed propagating frame widely available made in plastic. Various types are available from the simple covered box, consisting of a seedbox-sized base with a plastic dome, to the full thermostatically-controlled heated frame, with variable ventilators. The prices quite naturally rise with the degree of sophistication.

When purchasing any kind of propagator always make sure it has sufficient heating capacity for your requirements. A propagator for a cold greenhouse will need to have more in reserve than one in an already heated environment. With thermostatic control a greater heating capacity will cost no more to run than a less powerful unit. As a rule I would always advise buying the largest and best you can afford. Propagators, like most gardens, never seem large enough.

A note might be made here regarding mist propagation. This has led to a great improvement in the rooting potential of many previously 'difficult' cuttings, most notable among conifers and other evergreens. A mist unit is based on a specially constructed propagating frame containing under-bench heating controlled at round 21°C and mist irrigation (Fig. 46).

The mist is provided by fine nozzles set above the bed and these keep the plants continuously moist and the atmosphere around them at maximum humidity. The amount of mist is controlled by an electronic 'leaf' which senses when the atmosphere becomes dry and operates a solenoid valve which in turn turns on the mist until the necessary dampness is achieved.

Owing to the continuous warm, damp atmosphere, cuttings under mist do not wilt and require no shading. Under these conditions they continue to grow and soon form roots.

A free-draining compost is required, this usually consisting of silver sand alone or mixed with a small amount of peat. Cuttings are either rooted in this compost in a bed or in boxes which ideally are fitted with mesh bottoms to assist with drainage. There are no nutrients in the compost, and therefore plants rooted in this way must be transplanted as soon as roots are established. After

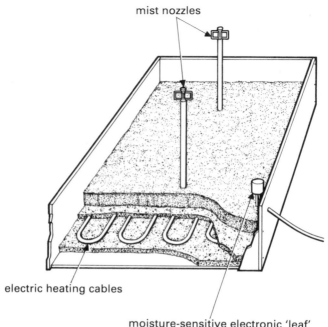

FIG. 46 Layout of a small mist propagation unit.

transplanting they should be kept in a shaded, close atmosphere for a time, gradually weaning them to an ordinary greenhouse environment.

Amateur units are available in kit form, or individual components can be bought.

The technique used in mist propagation is not new but merely mechanised. Nevertheless it can be duplicated by the dedicated gardener with a frame with bottom heat, the dampness being supplied by means of a syringe which is applied frequently in sunny weather and not so much in dull. Constant attention is necessary so it is not a method for those out at work all day.

All greenhouses, frames and propagators, especially those creating a warm damp atmosphere, should be kept as hygienic as possible by washing them down with disinfectant whenever possible. All disease and pests which do appear should be dealt with immediately one way or another.

Lastly we go back to the beginning with a house and a garden. Without any other facilities there is always the kitchen windowsill. A propagator of modest size does not look out of place there and neither does the home-made polythene bag propagator. This consists simply of a pot of compost in which cuttings or seeds are placed and a polythene bag cover which retains the necessary warm, close atmosphere provided by its larger cousins. The bag is simply popped over the top of the pot and held in place with a rubber band. Care should be taken to keep it away from any leaves by supporting it with either three short canes or a loop of stiff wire placed inside the pot. Some ventilation is necessary, especially when cuttings are just beginning to root, and is simply provided by snipping away the corners of the bag.

This method can also be used outside, preferably placing the pot with its cover in gentle shade under a bush or similar position until its mission is accomplished.

Index

Acanthus 21
Anemone coronaria 28
Anthemis 46
Aquilegia 28
Artemisia 46

Beans 27
Begonia 52−3
Begonia rex 17−19
Budding 78−87
   chip 85
   shield 74, 82−3
Budding knife 78, 79

Cacti 74, 75
Cambium layer 78
Camellia 12
Carnation 19, 57, 58
Carrot 26
Cedrus 75
'Chipping' 25
Chives 49
Chrysanthemum 13, 16
Clematis 16, 17, 60, 63
Cloches 88−9
Composts,
   John Innes Potting 36
   John Innes Seed 9
   Soilless 9
Cupressus 75
Cuttings 7−23
   basal 10, 11
   bud 10, 11−12
   eye 12
   half-ripe 10, 14−15
   hardwood 10, 15−16
   heel 10, 11, 12
   internodal 16, 17
   Irishman's 10, 16
   leaf 10, 17−19
   nodal 16
   piping 19−20
   root 20−21
   semi-hardwood 14−15
   softwood 10, 13−14
   stem 13−16
   tip 16

Dahlias 13, 52−3
'Damping off' 33
Division, 46−55
   fibrous roots 46−7
   offsets 48−50
   rhizomes 50
   suckers 50−51
   tubers 51−3
   woody roots 47−8
'Double working' 73−5

Eucalyptus 26

F1 hybrids 31−2
Ferns 40−42
Ficus 60
Forsythia 52
Frames 88−90

Gladiolus 48
Gloxinia 17
Gooseberries 61
Grafting 62−77
   approach 67−8
   bark 70
   cleft 64−5
   crown 71−3
   flat 75
   framework 68−71
   rind 71−3
   saddle 65−7
   splice 63
   stub 69
   veneer 75−6
   wedge 64−5
   whip and tongue 63

Hardening off 37
Helenium 46
Hemerocallis 52
Hormone rooting powder 8
Horseradish 20
Hybridisation 31−2

Ilex aquifolium 71
Iris (bearded) 50, 51
Ixia 49

Jasminum nudiflorum 11

Layering 56−61
  air 57−60
  serpentine 60
  tip 60−61
Lettuce 27
Lilium 49

Magnolia 26
Michaelmas daisies 46
Mist propagation 91−2
Mushrooms 42−3

Onions 27

Pears 73
Peas 27, 28
Perlargonium 7, 22
Phlox (HP types) 20
Picea 75
Pinus 75
Polyanthus 47
'Pricking out' 35
Primula 29, 31
Propagators 90−93
Pseudotsuga 75
Pteris cretica 'Wimsettii' 39

Raspberries 50, 55
Rhododendrons 66
Romneya coulteri 20
Rootstocks 62, 76−7, 85−7
Rosa canina 86, 87

Rosa rugosa 86, 87
Roses 12, 61, 63, 78
Rubber plants 60

Saintpaulia 17
Sansevieria 19
Scion 62−77
Seeds 24−45
  collecting 27−31
  dustlike 25
  fleshy 25, 26
  hard-coated 25
  oily 25, 26
  pelleted 24, 25
  plumed 25
  sowing 32−40
  winged 25, 26
Spawn 42−3
Spores 40−42
'Sports' 32
'Stem-building' 73
'Stratification' 25
Strawberry 56, 57
Sweet pea 25

Tomatoes 27
Tradescantia 14
Tulip 49

Vermiculite 9
Vines 12−13, 68

Weigela 15